CURRENT PROBLEMS IN FERTILITY

CURRENT PROBLEMS IN FERTILITY

Based on the IFA Symposium held in Stockholm, Sweden, April 2-4, 1970. Sponsored by Åhlen-stiftelsen, Sven och Dagmar Saléns stiftelse, and Roland Lundborg, M.D.

Edited by

Axel Ingelman-Sundberg

Professor of Obstetrics and Gynecology
Royal Caroline Institute
Chairman, Department of Obstetrics and Gynecology
Sabbatsberg Hospital, Stockholm

and

Nils-Olov Lunell

Assistant Professor of Obstetrical and Gynecological Endocrinology
Royal Caroline Institute
Sabbatsberg Hospital, Stockholm

PLENUM PRESS • NEW YORK-LONDON • 1971

Library of Congress Catalog Card Number 76-145405

SBN 306-30522-4

A Division of Plenum Publishing Corporation
227 West 17th Street, New York, N. Y. 10011

United Kingdom edition published by Plenum Press, London
A Division of Plenum Publishing Company, Ltd.
Donington House, 30 Norfolk Street, London W.C. 2, England

Printed in the United States of America.

FOREWORD

The tremendous increase in the output of scientific articles on reproduction is symptomatic of the growth of science within this field. Last year about 20.000 titles were listed. There is, consequently, a need for limited meetings, where actual problems in reproduction can be surveyed and discussed openly. In the IFA symposium in Stockholm 2-4 April 1970, 120 delegates from 18 different countries took part. The papers discussed on this occasion all have great general interest in the field of reproduction.

September, 1970

Axel Ingelman-Sundberg
President
International Fertility Association

CONTRIBUTORS

Jacob H. Aafjes, Department of Endocrinology, Growth and Reproduction, Rotterdam Medical Faculty, Rotterdam, The Netherlands

Matti Alanko, Department of Obstetrics and Gynecology, College of Veterinary Medicine, Hautjärvi, Finland

Nathan Amit, Semenology Institute, Tel Aviv, Israel

S. J. Behrman, Department of Obstetrics and Gynecology, University of Michigan, Ann Arbor, Michigan

G. Benagiano, I[a] Clinica Ostetrica e Ginecologica dell'Università di Roma, Rome, Italy

S. Bergström, Institute of Human Anatomy, Uppsala, Sweden

R. Blobel, University Frauenklinik, Tübingen, Germany

R. E. Bower, Department of Animal Science, University of Minnesota, St. Paul, Minnesota

G. Brenci, G. Mendel Institute for Medical Genetics and Twin Research, Rome, Italy

K. I. Brown, Department of Animal Science, University of Minnesota, St. Paul, Minnesota

L. Carenza, I[a] Clinica Ostetrica e Ginecologica dell'Università di Roma, Rome, Italy

Lars Carlborg, Department of Obstetrics and Gynecology, University Hospital, Uppsala, Sweden

B. G. Crabo, Department of Clinical Biochemistry, Royal Veterinary College, Stockholm, Sweden

R. G. Edwards, Physiological Laboratory, Cambridge University, United Kingdom

P. Ehrler, Viktoriaspital, Berne, Switzerland

Rune Eliasson, Department of Physiology, Faculty of Medicine, Karolinska Institutet, Stockholm, Sweden

M. Ermini, I[a] Clinica Ostetrica e Ginecologica dell'Università di Roma, Rome, Italy

A. Fabbrini, Istituto di Patologia Medica II, Università di Roma, Rome, Italy

George L. Foss, Male Subfertility and Endocrine Clinics, Bristol, England

J. Friberg, Department of Obstetrics and Gynecology, Institute of of Human Anatomy, Uppsala, Sweden

Mirjam Furuhjelm, The Department of Obstetrics and Gynecology, Karolinska Institutet, Sabbatsberg Hospital, Stockholm, Sweden

L. Gedda, G. Mendel Institute for Medical Genetics and Twin Research, Rome, Italy

E. F. Graham, Department of Animal Science, University of Minnesota, St. Paul, Minnesota

Carl G. Groth, Department of Surgery, Karolinska Institute, Serafimer Hospital, Stockholm, Sweden

I. Halbrecht, Department of Obstetrics and Gynecology, Hasharon Hospital, Petah Tiqua, Israel

Motoyuki Hayashi, Department of Obstetrics and Gynecology, Toho University, Tokyo, Japan

P. O. Hedlund, St. Erik's Hospital, Stockholm, Sweden

S. Heller, University Frauenklinik, Tübingen, Germany

Carl-Axel Ingemanson, Obstetrics and Gynecology Department, Central Hospital, Eskilstuna, Sweden

J. E. Kihlström, Institute of Zoophysiology, Uppsala, Sweden

Fred A. Kincl, The Population Council, Rockefeller University, New York

Masanori Kirimura, Department of Obstetrics and Gynecology, Toho University, Tokyo, Japan

L. Komlosh, The Research Laboratory of Human Reproduction and Fetal Development, Tel Aviv University Medical School, Tel Aviv, Israel

Argo Kövamees, Department of Surgery, Serafimer Hospital, Karolinska Institute, Stockholm, Sweden

Zdenek Kubes, Department of Gynecology and Obstetrics, New Jersey College of Medicine, New Jersey

E. Kučera, Second Department of Obstetrics and Gynecology, Charles University, Prague, Czechoslovakia

Stig Kullander, Department of Obstetrics and Gynecology at the Malmö General Hospital, Malmö, Sweden

Ulf Larsson-Cohn, Department of Obstetrics and Gynecology, University Hospital, Uppsala, Sweden

A. Lima-de-Faria, Institute of Molecular Cytogenetics, University of Lund, Lund, Sweden

Nils-Olov Lunell, Department of Obstetrics and Gynecology, Karolinska Institute, Sabbatsberg Hospital, Stockholm, Sweden

F. Macků, Second Department of Obstetrics and Gynecology, Charles University, Prague, Czechoslovakia

Gösta Magnusson, Surgical Department, Serafimerlasarettet, Stockholm, Sweden

Göran Möller, Department of Bacteriology, Karolinska Institutet, Stockholm, Sweden

Abraham Mor, Semenology Institute, Tel Aviv, Israel

O. Nilsson, Institute of Human Anatomy, Uppsala, Sweden

Jozi Ono, Department of Obstetrics and Gynecology, Toho University, Tokyo, Japan

S. Paatsama, Department II of Obstetrics and Gynecology, University Central Hospital, Helsinki, Finland

M. M. Pace, Department of Animal Science, University of Minnesota, St. Paul, Minnesota

Bengt Persson, S:t Göran Hospital, Karolinska Institute, Stockholm, Sweden

I. Petersén, Department of Clinical Neurophysiology, The Sahlgren Hospital, University of Gothenburg, Gothenburg, Sweden

M. Re, Instituto di Patologia Medica II, Universita di Roma, Rome, Italy

Harry W. Rudel, Biological Concepts, Inc., New York, New York

M. Savioli, Istituto di Patologia Medica II, Università di Roma, Rome, Italy

M. Seppälä, Department II of Gynecology and Obstetrics, University Central Hospital, Helsinki, Finland

L. Silvani, G. Mendel Institute for Medical Genetics and Twin Research, Rome, Italy

G. Spera, Istituto di Patologia Medica II, Università di Roma, Rome, Italy

P. C. Steptoe, Oldham and District General Hospital, Lancashire, England

R. Stěrbă, Research Institute of Pharmacy and Biochemistry, Prague, Czechoslovakia

B. Studený, Second Department of Obstetrics and Gynecology, Charles University, Prague, Czechoslovakia

A. D. Subrizi, I[a] Clinica Ostetrica e Ginecologica dell'Università di Roma, Rome, Italy

Thure Swensson, Uppsala läns Seminförening, Enköping, Sweden

R. Tachezy, Second Department of Obstetrics and Gynecology, Charles University, Prague, Czechoslovakia

G. Torrioli-Riggio, G. Mendel Institute for Medical Genetics and Twin Research, Rome, Italy

V. Trnka, Second Department of Obstetrics and Gynecology, Charles University, Prague, Czechoslovakia

P. Vara, Department II of Gynecology and Obstetrics, University Central Hospital, Helsinki, Finland

Jan Th. M. Vreeburg, Department of Endocrinology, Growth and Reproduction, Rotterdam Medical Faculty, Rotterdam, The Netherlands

Barbro Westerholm, Department of Pharmacology, Karolinska Institute, Stockholm, Sweden

CONTENTS

THE USE OF BLASTOCYSTS IN HUMAN AND VETERINARY MEDICINE

RECENT ADVANCES IN CONTRACEPTION

INSEMINATION IN HUMAN AND VETERINARY MEDICINE

CULTURE OF HUMAN MALE GERM CELLS IN VITRO

By A. Lima-de-Faria

Institute of Molecular Cytogenetics

Univ. of Lund, Tornavägen 13, Lund, Sweden

A human individual results from the fertilization of an egg by a spermatozoon. Both cells carry a haploid number of human chromosomes (n=23) and the genetic information that determines the differentiation of the cells into an individual is constituted by the deoxyribonucleic acid (DNA) located in the chromosomes. The genetic constitution of a human being can only be changed if its DNA is changed. The DNA that builds the chromosomes replicates at a given interval during the cell cycle, and it is at this time that a new DNA chain, and subsequently a new chromosome, are formed.

Thus, if we are going to interfere with the genetic material of man it will be easier to do it: (1) at the time new DNA is being synthetized and (2) at the moment of formation of the chromosome which is going to become part of the spermatozoon.

Women are born with their eggs at diplotene (the end of prophase of meiosis). This means that DNA synthesis has occured in the embryonic stage which makes it unaccessible to experimental control. In men, on the other hand, the DNA synthesis that leads to the formation of chromosomes which become part of the sperm, occurs in the cells of adult testes. If those cells could be cultivated *in vitro*, the possibility would be open for interfering with the human genetic material in a more direct form.

We have recently been able to develop a method that allows the *in vitro* culture of human germ cells from the time of DNA synthesis to late pachytene, i.e. for a period of 16 days. This is about one quarter of the time between DNA synthesis and the appearance of mature spermatozoa (64 days).

By exposing the germ cells to H^3-thymidine, which is selectively incorporated into DNA during its replication, it is possible to establish two facts. First, the germ cells undergo DNA synthesis *in vitro*. Secondly, since the chromosomes labelled can be identified by autoradiography, it can be ascertained which cells developed in culture from the time of DNA synthesis.

The results presented were published previously: A. Lima-de-Faria, J. German, M. Ghatnekar, J. McGovern and L. Andersson, 1968, Hereditas 60:249-261. The report that follows is a condensation of the methods and results described in that paper.

Materials and methods

In this study 14 patients aged 26-79 were used: 6 from Cornell University Medical College, New York, and 8 from the University Hospital at Lund. In most cases the biopsies were obtained during prostatectomy.

This investigation consisted of two types of experiments. In the initial study biopsies from 5 patients were used. These were placed in culture medium containing H^3-thymidine and incubated at 37°C for periods varying from 30 minutes to 28 hours. The material was fixed in acetic-alcohol 1:4 immediately after exposure to the isotope.

In the second type of experiment biopsies from 9 patients were pulse labelled for periods of time varying from 30 to 120 minutes. The isotope was removed and the cells were throughly washed in three changes of fresh medium without H^3-thymidine. Subsequently they were incubated at 37°C for a period of time ranging from 5 to 16 days, during which the specimen was transferred to fresh medium every 2 to 3 days. When the culture was terminated the cells were fixed in acetic-alcohol 1:4.

In both types of experiments the small specimens were minced with scissors and were incubated in culture medium containing 25 μc/ml or 50 μc/ml H^3-thymidine (specific activity 1.9 c/mM). Two culture media were used: (1) PARKER'S 199 containing 15 % calf serum and (2) EAGLE'S MEM containing 10 % calf serum, each with penicillin and streptomycin.

Sections (3μ) and squashes of the material were stained with the Feulgen reaction. Autoradiographs were prepared using Kodak AR-10 stripping film. The exposure to the isotope varied from 8 to 26 days. Heavily labelled cells were photographed. Subsequently the silver grains were removed, in order to identify the labelled cells, using the technique of BIANCHI et al. (1964).

RESULTS

In the first type of experiment involving exposure to H^3-thymidine followed immediately by fixation, many cells appeared heavily labelled in the sections as well as in the squashes. Sections were used in this experiment mainly because this allows one to locate the cells in the seminiferous tubule. This helps in determining the developmental stage of the spermatocytes since there is a wave of meiosis which develops from the peripheral cell layers of the seminiferous tubule - where spermatogonia are located - and moves towards the centre.

The sections showed very heavily labelled cells mainly in the outer layers of the seminiferous tubules (Fig.1). When the silver grains were removed these cells were seen to be at the interphase preceding meiosis and at leptotene (Fig.2). Leptptene in the human male is characterised by a very dense juxtaposition of the chromosomes resulting in compact and heavily stained nuclei. In this patient (No. 2) the cells were exposed for 1 hour and 30 minutes to tritiated thymidine (Table 1). A similar result was obtained in the following patients.

In patient No. 4 after 7 hours and 30 minutes of exposure to tritium many cells were found labelled. In Fig. 3 there are four cells heavily labelled. After removal of the silver grains it may be seen that three of these cells are at leptotene-zygotene. They show the typical thread like chromosomes with small chromomeres and the whole nucleus has the compact appearance characteristic of leptotene. Protruding from the main nuclear mass can be seen single chromosomes and chromosomes with double the thickness disclosing the beginning of pairing characteristic of zygotene (Fig. 4). The same type of cells were found labelled in patient No. 5 after three hours of exposure to tritium. In cells from patient No. 1 in which the exposure to labelled thymidine was only 30 minutes one finds besides labelled leptotene nuclei, labelled cells which are at leptotene-zygotene, if not at zygotene (Figs. 5 and 6).

TABLE 1. Labelling of meiotic cells in man after exposure to H^3-thymidine in vitro.

Patient (reference number)	Age	Time in H^3-thymidine	Time between H^3-thymidine and fixation	Main stages labelled
1	71	30 min.	0	Interphase, leptotene, leptotene-zygotene
2	57	1 hour 30 min.	0	Interphase, leptotene
5	53	3 hours	0	Leptotene, zygotene
4	70	7 hours 30 min.	0	Leptotene, zygotene
13	65	2 hours	5 days	Early and mid-pachytene
13	65	2 hours	10 days	Mid-pachytene
6	76	30 min.	14 days	Mid-pachytene and late pachytene
14	34	45 min.	14 days	Very late pachytene
7	76	35 min.	15 days	Mid-pachytene
8	66	35 min.	15 days	Mid-pachytene
11	79	35 min.	16 days	Mid-pachytene

These results are summarised in the first part of Table 1 and disclose that, under the conditions of our experiments, DNA synthesis not only takes place at the interphase preceding meiosis, but also occurs at the early stages of prophase: leptotene and leptotene-zygotene.

The second type of experiment, in which the cells were pulse labelled with H^3-thymidine, thoroughly washed, changed to fresh medium without isotope and allowed to develop for a period of time extending throughout 16 days, revealed that the meiotic cells apparently developed normally in the culture medium during prophase of meiosis.

Early and mid-pachytene cells started to appear heavily labelled after five days in culture. After 10 days in culture labelled mid-pachytene cells were mainly found (Table 1, patient 13). At mid-pachytene the chromosomes are well individualised with their chromomere structure quite distinct, but the bivalents are still quite long and get entangled into each other, being quite difficult to separate on a squash. The sex vesicle is not yet distinct (Figs. 7 and 8). No difference can be seen between the structure of these chromosomes cultured in vitro and those fixed from cells growing in vivo.

After fourteen days in culture the spermatocytes of patient No. 6 (76 years old) showed mid-pachytene and late pachytene nuclei heavily labelled (Figs. 9 and 10). At this stage the sex vesicle is quite distinct and the autosome bivalents have now become much shorter. These are cells which were labelled at interphase of meiosis or early prophase and took subsequently 14 days to develop into late pachytene. In every respect they look normal. It is known from the work of HELLER and CLERMONT (1963) that the spermatocytes of man take 16 days to develop in vivo from interphase of meiosis to mid-pachytene. Thus, there is very good agreement between the rate of development in vitro in our cultures as compared to in vivo conditions.

Tissue derived from a younger patient (No. 14), aged 34 years, showed labelled cells at a still more advanced stage, after 14 days in culture. Large numbers of very late pachytene cells, where most bivalents could be easily separated by the squash technique, were found heavily labelled. Three other older patients (ages 76, 66 and 79) had cells that developed well throughout 15 and 16 days. The labelled cells in the three cases reached mid-pachytene (Table 1).

Most specimens survived quite well in culture. Of the nine patients used in the second experiment, only in three cases did the cells develop less well with the subsequent arrest of meiotic prophase at an early stage. This may have been due to the pathological condition of the patient or to the size of the specimens cultured. We have found that specimens cut into one cubic millimeter of tissue are prone to develop better than larger ones.

Note - For the Figs. mentioned in the text see original paper in "Hereditas".

DISCUSSION

The results just described have a bearing on two problems: (1) The period of DNA synthesis related to meiosis, and (2) the culture of human male germ cells under defined conditions. In this report we will only discuss this second problem.

It is known from the work of HELLER and CLERMONT (1963) that it takes 16 days for the human male germ cells to develop from preleptotene to mid-pachytene. The development time from interphase to immature spermatids and free spermatozoa is approximately 32 and 64 days respectively.

Our results (Table 1) show that the human cells in culture develop as fast as in the human body. Moreover, the appearance of the cells and of the chromosomes is indistinguishable from normal material. Thus, it is possible to cultivate human germ cells throughout most of the prophase of meiosis under controlled conditions.

How far one will be able to obtain normal development of these cells throughout the whole process of meiosis, to spermatids and spermatozoa is to be found out. We know, however, from previous work on the culture of meiotic cells in plant material (LIMA-DE-FARIA, 1950) that once the cells are placed in culture at the beginning of meiosis they proceed through all meiotic stages without difficulty. In animals the formation of spermatids and spermatozoa is known to lead to ultrastructural and histochemical changes in the chromosomes. Thus, at spermatid and spermatozoa formation one will propably encounter various difficulties, but these may not be impossible to surmount.

SUMMARY

(1) Testicular biopsies were obtained from 14 patients aged 26-79. In most cases the biopsies were obtained during prostatectomy. The small specimens were minced with scissors and were subsequently incubated at 37°C in culture medium containing 25 μc/ml or 50 μc/ml H^3-thymidine (specific activity 1.9 c/mM). Two culture media were used: (1) Parker's 199 containing 15 % calf serum and (2) Eagle's MEM containing 10 % calf serum, each with penicillin and streptomycin. Two types of experiments were performed.

(2) In the first type of experiment the cells were continuously exposed to H^3-thymidine for periods of time

ranging from 30 minutes to 28 hours. In each case the cells were immersed in acetic-alcohol 1:4 immediately after exposure to the isotope. Sections (3 μ) were stained according to the Feulgen procedure and autoradiographs were prepared using Kodak AR 10 stripping film. The exposure of the film to the isotope varied from 8 to 26 days. Heavily labelled cells were photographed. Subsequently the silver grains were removed in order to identify the labelled cells. In specimens exposed to H^3-thymidine for 90 minutes, labelling was found in nuclei of cells at the interphase preceding meiosis and at leptotene. Interphase, leptotene, and leptotene-zygotene nuclei were heavily labelled in cells fixed after 30 minutes of exposure to H^3-thymidine. These results indicate that under the conditions of the present experiments DNA synthesis in human spermatogenic cells takes place during the interphase preceding meiosis and also occurs at the leptotene and leptotene-zygotene stages.

(3) In a second type of experiment the cells were exposed to H^3-thymidine for 30 to 120 minutes, the isotope was removed and the cells were thoroughly washed in 3 changes of fresh medium. Subsequently they were incubated at 37°C for a period of time ranging from 5 to 16 days during which the specimen was transferred to fresh medium every 2 to 3 days. Feulgen squashes were prepared. In the tissues fixed after 5 days in culture early and mid-pachytene cells were labelled, and in cells fixed after 14 days late pachytene nuclei were shown to be heavily labelled. Human meiotic cells apparently develop normally in vitro from the period of DNA synthesis preceding meiosis to near the end of the prophase of the first meiotic division. The in vitro incorporation of H^3-thymidine into human chromosomes of cells that may lead to the formation of spermatozoa opens the possibility of interfering with the genetic material of man in a readily accessible form.

LITERATURE CITED

BIANCHI, N., LIMA-DE-FARIA, A. and JAWORSKA, H. 1964. A technique for removing silver grains and gelatin from tritium autoradiographs of human chromosomes. - Hereditas 51:207-211.

HELLER, C.G. and CLERMONT, Y. 1963. Spermatogenesis in man: an estimate of its duration. - Science 140:184-186.

LIMA-DE-FARIA, A. 1950. Meiosis and pollen mitosis in rye under controlled conditions. - Hereditas 36:106-109.

PRESERVATION OF HUMAN SPERM BY LIQUID NITROGEN VAPOR FREEZING

S. J. Behrman, M. D.

Professor, Department of Obstetrics and Gynecology

University of Michigan, Ann Arbor,Mich.

The ability to freeze preserve human sperm by cryogenic methods has several very powerful advantages: (1). The availability of sperm for heterologous insemination independent of time and place (2) the collection and centrifugation of oligospermic specimens for homologous insemination (3) for preservation of sperm prior to vasectomy to enhance population control (4) for selective breeding (5) for the stock-piling of national resources much as we do other assets, and (6) for the study of normal physiology. Our technique has been previously described[1,2] and need not be repeated here. However, I would like to present some of the problems encountered in this form of research both at a basic physiologic level and with its resultant clinical implications. Our method of cooling (Figure 1) and thawing (Figure 2) are collated in (Figure 3) in relationship to time and temperature. As of this time the pregnancy rate with frozen sperm in heterologous insemination is 48.2% as compared to 68% with fresh human sperm requiring 6.9 inseminations on the average over a 4.2 month period. Of patients that became pregnant 93% did so in the first six months with frozen sperm as compared to 92.8% when using fresh sperm.

Of great interest, however, is the finding that the post-thaw motility of frozen preserved sperm is not a determinant or index of its subsequent fertility. It was this finding that led to our belief that motility per se is firstly a poor criterion for the determina tion of a successful donor specimen; and secondly that the damage entailed in the freeze preservation technique must be examined in the light of other parameters than motility. It was strongly hypothesized that metabolic failure or irreversible alteration is perhaps responsible for fertility differentials in frozen preserved sperm as compared to fresh sperm. In line with this hypothesis a study[3]

Table I

MOTILITY AND RESPIRATION

A. Between Room Temperature, 22°C and 0°C 6 Cases.....Mean of Sperm Motility.....64% (3.8)

	% Motility (Mode of Motility) After Cooling	After Incubation	$2O_2$ / 4 hours (µl / 10^8 sperm)
Control	64% (3.7)	49% (3.0)	51.9 µl
Rapidly Cooled 15°C-20°C / min.	61% (3.5)	40% (2.0)	37.1 µl
Slowly Cooled 1°C / min.	64% (3.8)	45% (2.6)	49.0 µl

B. At the Range of Heat Diffusion 4 Cases.....69% (4.0)

Rapidly Passed	56% (4.0)	35% (3.0)	31.1 µl
Slowly Passed	49% (3.8)	28% (2.8)	22.4 µl

C. Between 0°C (after heat diffusion) and -30°C 6 Cases.....60% (4.0)

Rapid 20°C / min.	35% (3.5)	25% (2.8)	20.1 µl
Medium 5-7°C / min.	28% (3.5)	22% (2.7)	27.2 µl
Slow 2°C / min.	26% (3.0)	20% (2.7)	24.0 µl

D. Below -30°C to -85°C 5 Cases..... 55% (3.8)

Rapid 50°C / min.	33% (3.2)	29% (2.2)	23.5 µl
Slow 7 - 10°C / min.	34% (3.4)	25% (2.2)	23.2 µl

correlating rates of cooling temperature changes with post-thaw motility and respiration of sperm, between room temperature and -80° was undertaken. It was found that rapid cooling of semen above the freezing point, i.e. in the range of temperature shock, reduced the motility, but reduction in respiratory activity was much more severe. A rapid rate of cooling at the range of heat of fusion was preferable with respect both to motility and respiration, to slower rates of cooling in this temperature range. No further damage to either respiration or motility occurred at temperatures below -30°C. In general, slower cooling methods and slower thawing methods seemed preferable to faster ones. It appeared that the significant damages occurring to sperm in the range of temperature shock, between room temperature and the initial ice crystallization.[4]

It was also demonstrated that temperature shock and frozen specimens incubated anaerobically produced more lactate than do specimens which underwent other combinations of cooling and gas conditions. Altered molar relations between fructose utilization and lactate production in the presence of oxygen indicate that the most pronounced effect of cooling upon sperm occurred in the range of temperature shock as would have been predicted on the previous findings. Cooling in the range of thermal shock, room temperature to 5°C, has the effect of inhibiting fructolysis in human spermatozoa and this inhibition is independent of cellular death. This indicates a pronounced alteration of cell physiology which is independent of motility. This probably has the effect of making sperm metabolism increasingly glycolytic and decreasingly oxidative. This in turn means that as a result of cooling, sperm probably glycolyses glycerol rather than fructose. One would assume that this metabolic alteration has the ultimate effect of being toxic to the spermatozoa. If this is true it would account for the fact that frozen sperm must be inseminated within twenty-four hours of ovulation in order for pregnancy to occur as has been indicated by our clinical data[2]. Furthermore, it also appears that penetration of sperm after being frozen through cervical mucus by in vitro testing[5] shows a marked reduction of ability of the sperm to penetrate through cervical mucus.

On the basis of post-thaw motility studies it was found that sperm survival decreases linearly as a function of storage at warming temperatures between the temperatures of -100°C and -50°C. It is the rising storage temperatures rather than the original temperatures at which the specimen is stored which determines the motility loss. Damage is most extensive during storage at warming temperatures between -75°C and -50°C. This may indicate that the recrystallization at warming temperatures in storage may contribute to low temperature damage.

In summary, the work to date in our laboratory indicates two major findings: Firstly, that the significant damage occurring to

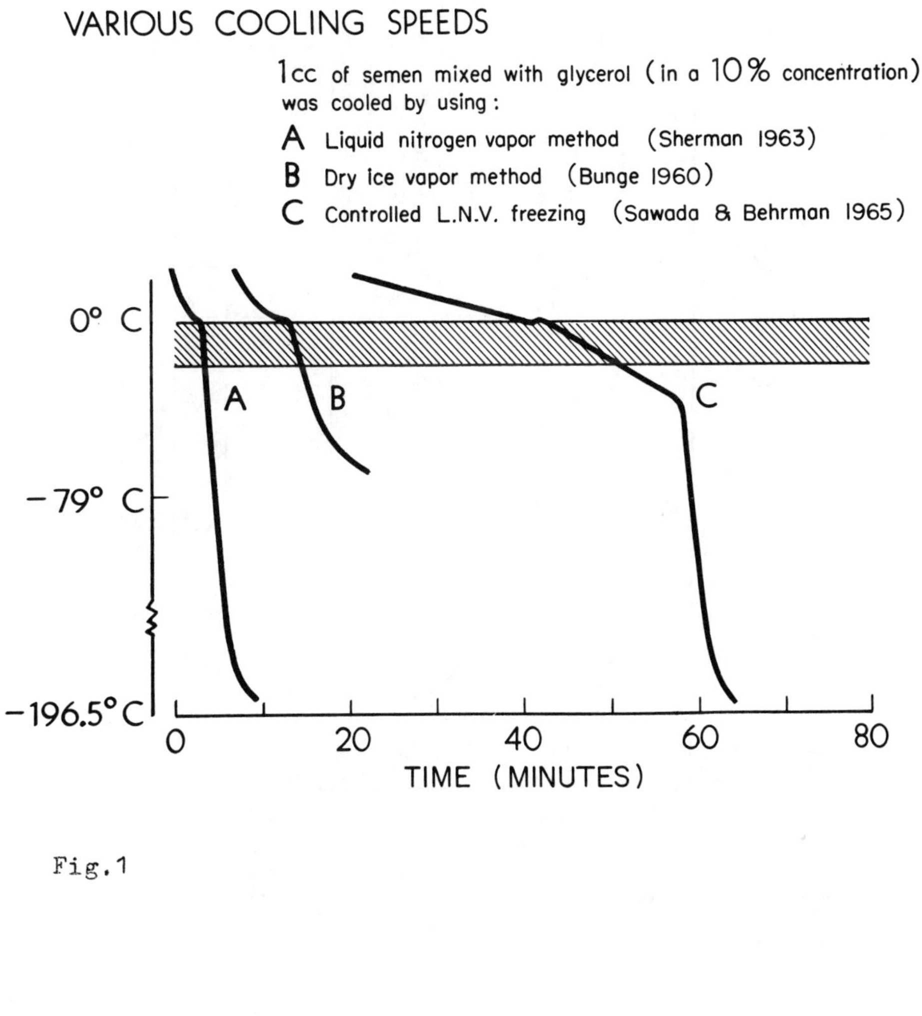
VARIOUS COOLING SPEEDS
1cc of semen mixed with glycerol (in a 10% concentration) was cooled by using:
A Liquid nitrogen vapor method (Sherman 1963)
B Dry ice vapor method (Bunge 1960)
C Controlled L.N.V. freezing (Sawada & Behrman 1965)
0° C
−79° C
−196.5° C
A
B
C
0
20
40
60
80
TIME (MINUTES)

Fig.1

VARIOUS THAWING METHODS

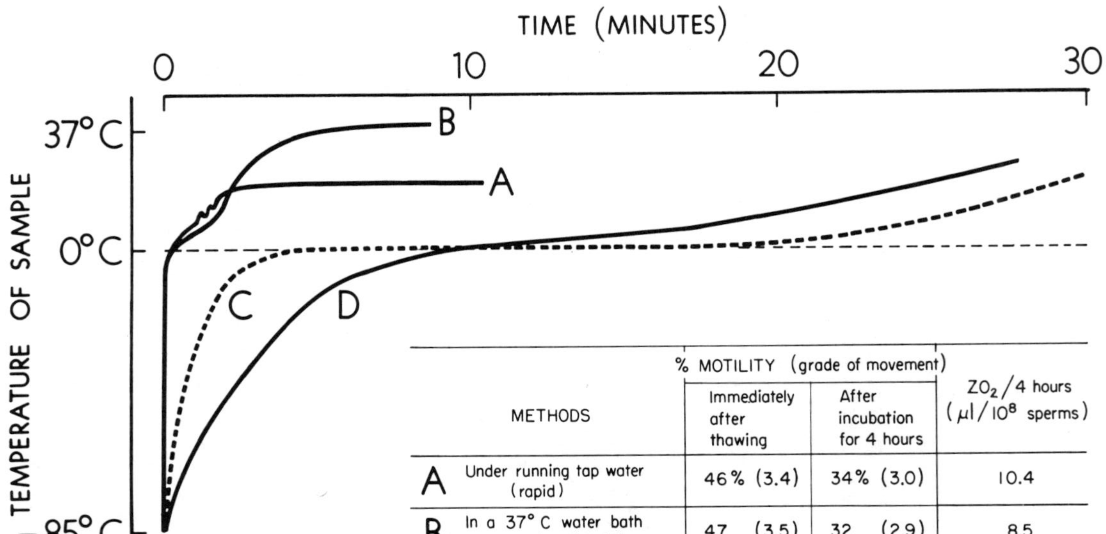

	METHODS	% MOTILITY (grade of movement) Immediately after thawing	After incubation for 4 hours	ZO_2/4 hours ($\mu l/10^8$ sperms)
A	Under running tap water (rapid)	46% (3.4)	34% (3.0)	10.4
B	In a 37° C water bath for 5 min. (rapid)	47 (3.5)	32 (2.9)	8.5
C	In a 5° C alcohol bath for 15 min. (med.-slow)	47 (2.8)	37 (2.8)	12.2
D	In an air at room temp. (slow) (22° C)	42 (3.0)	37 (2.8)	16.9

Fig.2

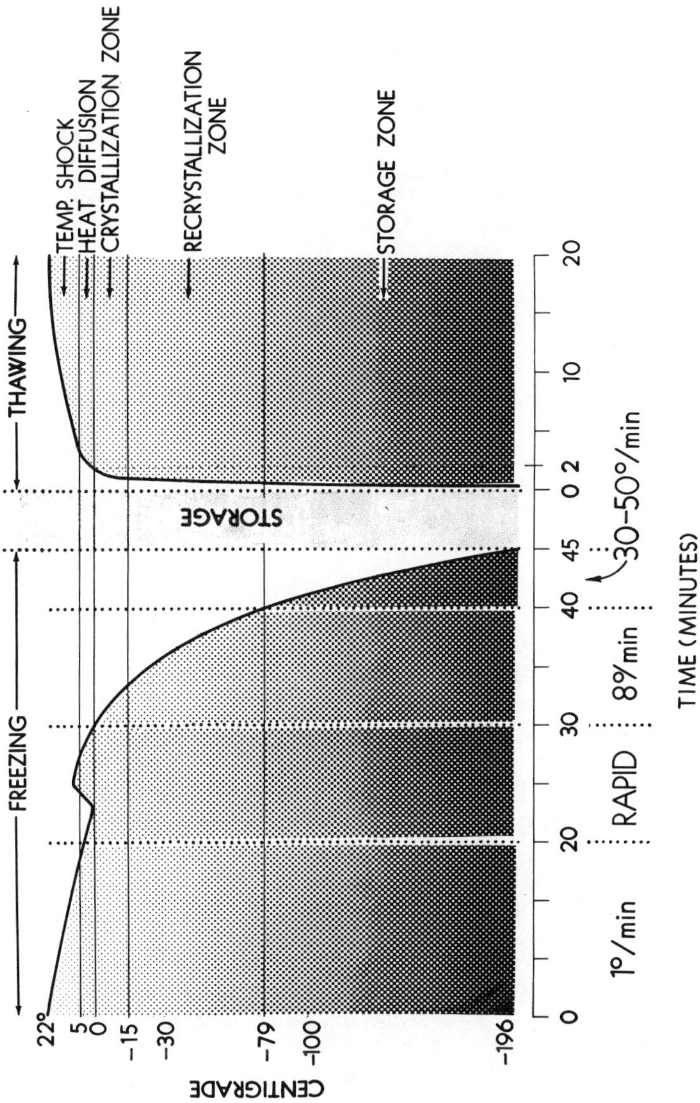
FREEZING
THAWING
STORAGE
TEMP. SHOCK
HEAT DIFFUSION
CRYSTALLIZATION ZONE
RECRYSTALLIZATION ZONE
STORAGE ZONE
CENTIGRADE
22°
5
0
−15
−30
−79
−100
−196
0
20
30
40
45
0 2
10
20
1°/min
RAPID
8°/min
30–50°/min
TIME (MINUTES)

Fig.3

these sperm cells as a result of freezing preservation are physiological and are independent of motility or cellular death during freezing. Secondly, that perhaps we can expect sperm cells to respond to the freezing preservation technique as a function of the genotype of the sperm donor, and that the genotype of the recipient as well may influence the fertilizing capacity of the cells so treated.

Finally, it is absolutely necessary to realize the very large difference existing between species with respect to ability of spermatozoa to survive freeze preservation. Thus, it is mandatory that we begin to formulate questions in terms of the cryogenic requirements of the human male gamete rather than in the terms of mammalian gametes in general. There are three areas of deficiency which must be answered immediately:

(1) How can semen be frozen and thawed in a way that is consistent with a very high fertilizing potential?

(2) What will be the biologic consequences to the human species of the application of this technique?

(3) What will be the sociologic consequences of the application of cryogenic techniques at any given frequency level of application in human population?

References

1. Behrman, S.J. and Sawada, Yoshiaki: Heterologous and homologous inseminations with human semen frozen and stored in a liquid-nitrogen refrigerator. Fertil.Steril.17(4) July-Aug.1966.
2. Behrman, S. J., and Ackerman, D. R.: Freeze preservation of human sperm. Amer. J. Ob. Gynec. 103 (5) 654-661, 1969.
3. Sawada, Y., Ackerman, D. and Behrman, S. J.: Motility and respiration of human spermatozoa after cooling to various low temperatures. Fertil. Steril. 18 (6) Nov.-Dec. 1967.
4. Ackerman, Donald R.: The effect of cooling and freezing on the aerobic and anaerobic lactic acid production of human semen. Fertil. Steril. 19 (1) Jan.-Feb. 1968.
5. Fjallbrant, B. and Ackerman, D.R.: Cervical mucus penetration in vitro by fresh and frozen-preserved human semen specimens. J. Reprod. Fert. 20, 515-517, 1969.
6. Ackerman, D. R.: Damage to human spermatozoa during storage at warming temperatures. Internat. J. of Fertil. 13 (3) 220-225, July-Sept. 1968.

MOTILITY AND MORPHOLOGY OF HUMAN SPERMS AFTER FEEZING IN LIQUID NITROGEN

Friberg, J. and Nilsson, O.

Department of Obstetrics & Gynecology

and Institute of Human Anatomy, Uppsala, Sweden

Freezing of human sperms without a protective medium destroys most of the sperms. However, if a protective medium is added to the sperm sample before freezing, the viability after freezing and thawing will be better. Many different substances have been used as protective agents. In this investigation the effect of glycerol and glycerol-egg-yolk-citrate on sperm freezing has been studied.

In the method of Sherman (4), sperms and glycerol are mixed in the proportion of 10 to 1. This mixture is put into plastic paillettes (Societe Anonyme, Instruments Med.vet., L'Aigle, France) which are placed in the vapours of liquid nitrogen. After 8 minutes in the vapours, the paillettes are lowered into the liquid nitrogen (the direct method).

In the method of Behrman-Sawada (1), sperms and glycerol-egg-yolk-citrate are mixed in the proportion of 1 to 1. A modification of this freezing method is used (2,3) in which paillettes ready to freeze are placed in +4° C for 25 minutes and then moved to a vessel containing alcohol at a temperature of -40° C. The paillettes are left at this temperature for 5 minutes and are then put directly into liquid nitrogen (the gradual method).

The lowering of the temperature during freezing by the two methods can be seen in Fig.1. The temperature is registered with a thermo-element inside paillettes frozen in a vertical position.

By Sherman's method the sperm sample is supercooled to -15° C. As the sample starts to freeze the temperature rises to -8 °C and then falls very quickly. In the lower part of the paillette the temperature goes down to about -180° C and in the upper part to about -110°C.

By Behrman-Sawada's method the temperature goes down slowly to +4°C. Supercooling and freezing show the same pattern as in Sherman's method but take a little longer time. When the sample is transferred to liquid nitrogen a small rise in the temperature can be seen. This rise occurs because the paillettes have to be moved in room temperature from the -40°C alcohol bath to the liquid nitrogen.

The temperature rise during thawing of the sperm samples is shown in Fig.2. Around -8° C the curve levels out for about 1 1/2 minutes after which the temperature continues to rise. During this plateau phase the sperm sample thaws.

In the present experiments, the motility of donor sperm samples was investigated in a fresh sample and after freezing according to methods I, II, III and IV in Fig.3. These methods included that of Sherman, Behrman-Sawada and in addition two modified methods in which glycerol was substituted in Behrman-Sawada's method and glycerol-egg-yolk-citrate in Sherman's method. The freezing and thawing temperature curves showed the same course in the two modified methods as in the original methods.

The modility was studied after thawing at room temperature and the activity was followed for 12 hours. Three portions of the fresh, pre-frozen samples, were used as controls, and three portions from each freezing method was investigated.

Two different sperm samples with different motility patterns are demonstrated in Figs. 4 and 5. The motility of the sperm sample of Fig. 4 has been reduced about 50% whereas the sperm sample in Fig. 5 shows a motility loss of about 20%. This is a common finding and it implies that the motility of the thawed sperms varies with different donors. For a particular donor, one of the tested methods sometimes resulted in better motility than the others. When donor samples, which had been treated by these four methods, were examined weekly during a two months period the motility pattern for a certain donor was observed to be rather constant. Thus, the method giving the best motility one week will also give the best motility during following weeks. Another point of interest is that the protective medium seems to be of greater importance than the freezing method. Thus, for instance, if glycerol in the direct method offers the best motility, glycerol in the gradual method will result in a better motility than the glycerol-egg-yolk-citrate methods.

The cervical mucus has been examined for surviving sperms 24 hours after the insemination. Sperms frozen according to Sherman and Behrman-Sawada were used for the inseminations and the women were inseminated during two subsequent ovulatory periods. Inseminations during the first ovulatory period were done with sperms frozen according to Sherman and inseminations during the second period

1

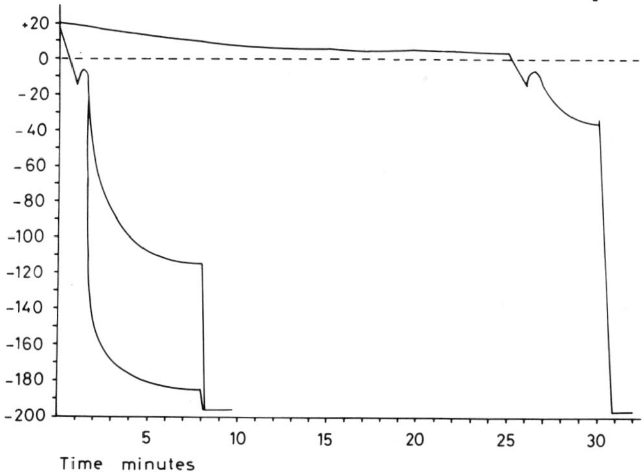

2

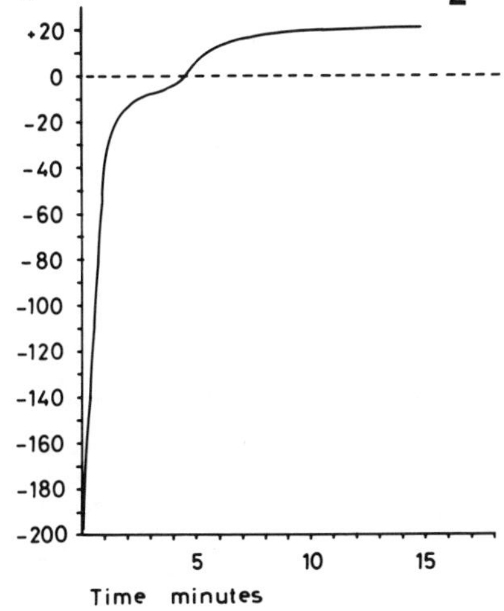

3

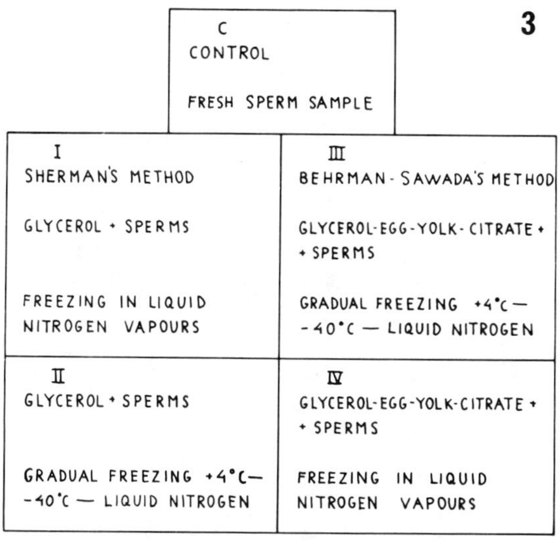

Fig. 1. Temperature recordings during freezing of human sperms with Sherman's and Behrman-Sawada's methods. Sherman's method is shown to the left where the upper curve is the temperature in the part of the paillette which is placed away from the liquid nitrogen, the lower curve is the temperature in the paillette just above the surface of the liquid nitrogen. To the right is shown the temperature when Behrman-Sawada's method has been used.

Fig. 2. The temperature rise during thawing of the sperm sample in room temperature. The same curve was seen with both Sherman's and with Behrman-Sawada's methods.

Fig. 3. The different freezing methods used in the motility investigations of a donor sperm sample.

4

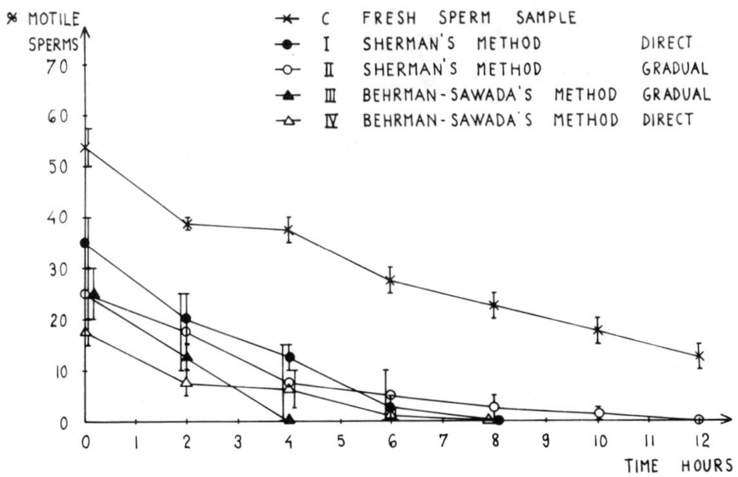

5

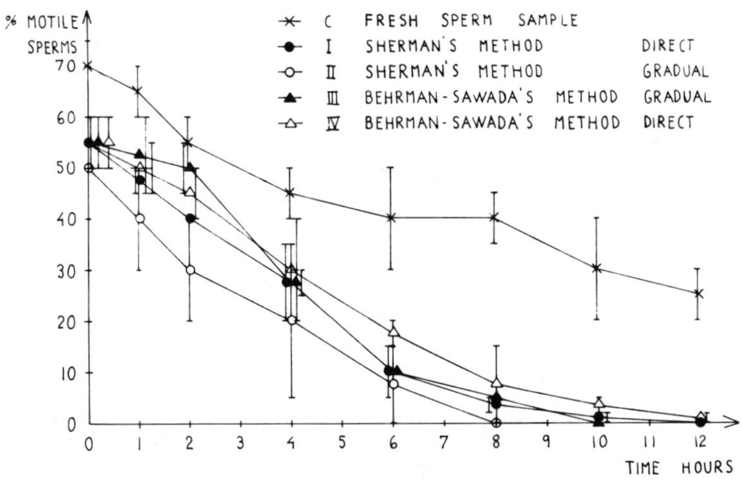

Fig. 4 and 5. The motility of two different donor sperm samples followed for 12 hours. The motility of the control portion is seen in the upper part of the figures and that of the portions obtained by the different freezing methods in the lower part.

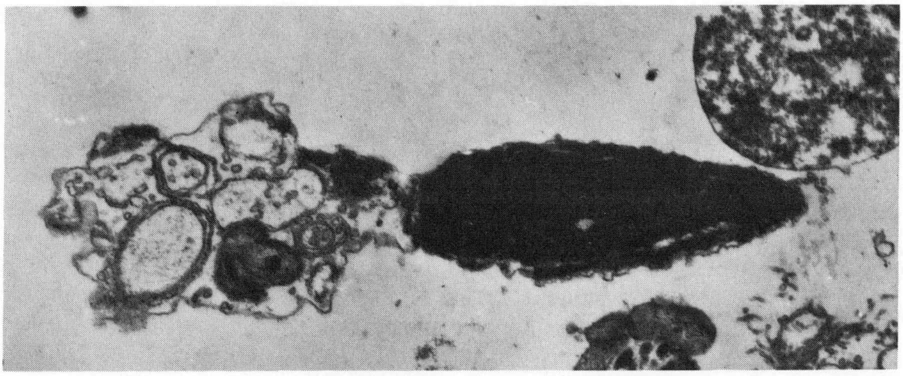

6

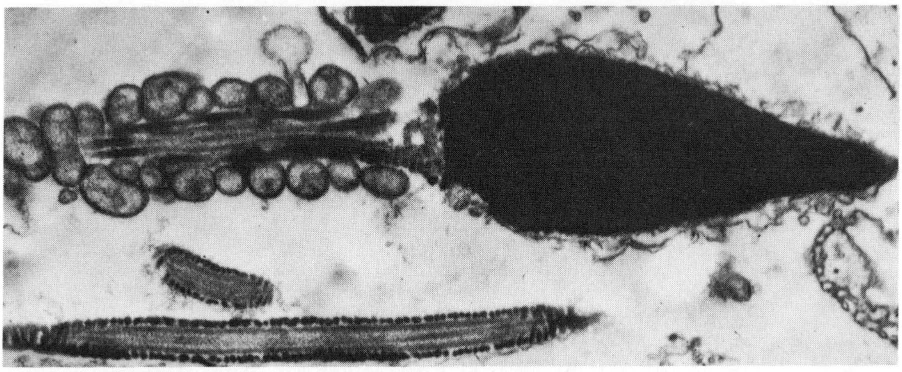

7

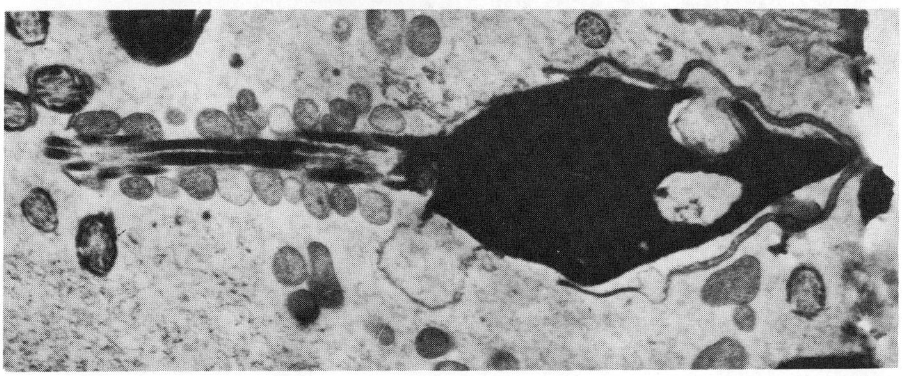

8

Fig. 6. Human sperm frozen without any protective medium. Note the destruction of the middle piece. Mag. 18 000 X.

Fig. 7. Human sperm frozen according to Sherman with glycerol as the protective medium. Mag. 18 000 X.

Fig. 8. Human sperm frozen according to Behrman-Sawada. Mag. 18 000 X.

were done with sperms frozen according to Behrman-Sawada. In the preliminary study, more surviving sperms were found when sperms frozen with Sherman's method were used than when sperms frozen with Behrman-Sawada's method were used.

The electron microscopical appearance of human sperms after freezing and thawing differs with the two freezing methods. When no protective medium was used the most significant damage was observed in the middle piece where the mitochondria were swollen and in some cases ruptured. The acrosome was also noted to be ruptured and wrinkled (Fig. 6). Sperms frozen without a protective medium can be seen to move the tail piece slowly but they do not have any active propagation. This can probably be explained by the destruction of the mitochondria.

Freezing with a protective medium, however, results in mitochondria that retained most of their normal morphology (Figs. 7 and 8) and the motility of the sperms was also rather good. With glycerol as a protective medium (Fig. 7) the acrosome is much more wrinkled than when glycerol-egg-yolk-citrate has been used as a protective medium (Fig. 8). Thus, in the electron microscope sperms frozen according to Behrman-Sawada show less damage than sperms frozen according to Sherman.

References

1. Behrman,S.J., and Sawada,Y.: Heterologous and homologous inseminations with human semen frozen and stored in a liquid-nitrogen refrigerator. Fert. & Steril.17: 457-466, 1966.

2. Matheson,G.W., Carlborg,L., and Gemzell,C.: Frozen human semen for artificial insemination. Am.J.Obstet.& Gynecol.,104: 495-501, 1969.

3. Sawada, Y., and Ackerman,D.R.: Use of frozen human semen. In Behrman,S.J., and Kistner,R.W., editors: Progress in Infertility, Boston, Little, Brown & Company, 1968.

4. Sherman,J.K.: Improved methods of preservation of human spermatozoa by freezing and freeze-drying. Fert. & Steril.14: 49-64, 1963.

This work was supported by the Ford Foundation (Grant No 66--405 and a Post Doctoral Fellowship to J.F.) and by the Swedish Medical Research Council (Grant No B70-12X-70).

SOME PROBLEMS INVOLVED IN FREEZING AND INSEMINATION WITH HUMAN SPERM

Lars Carlborg

Department of Obstetrics and Gynaecology

University Hospital, Uppsala, Sweden

Modern cryobiology has made possible storing of sperm at low temperatures with retained post thaw fertilizing capacity. The incidence of pregnancies in the human has unequivocally been reported to be lower when frozen sperm is used compared to the use of fresh sperm. The excellent results which have been obtained in veterinary medicine using frozen bovine sperm have not yet been realised in human medicine and will necessitate further research on human sperm. Results obtained from animal research will be only of limited value.

Some of the difficulties met with in this field of research are the assessment of fertile sperm. The ultimate answer to this question is of course the outcome of pregnancy, but large scale experiments using a particular research batch of frozen semen, as is done when making bovine experiments, cannot be performed in the human. The scientist will mainly have to resort to laboratory data. Both when dealing with fresh or frozen sperm, available laboratory methods will give only crude information on the probable fertility of the sperm samples. With regard to frozen and thawed sperm, comparisons can be made between data obtained before the freezing procedure and after thawing. Most investigators compare in this way the difference in motility, quantitatively and qualitatively. We have frequently observed that this procedure is of questionable value, and no certain correlation exists between good motility and subsequent fertilization. Insemination with thawed samples with a comparatively low qualitative motility may result in a higher incidence of pregnancy than a sperm sample with a better motility. In addition, we have also observed that frozen sperm of certain donors will give a higher pregnancy rate after insemination than other sperm with equally good routine data. It is assumed that newer methods based upon oxygen comsumption of spermatozoa or the content of enzymes will provide better means in the assessment of fertility. Possible damage caused by the freezing procedure under investigation may be detected using electron microscopy.

When freezing a sperm sample the problems are concerned with the rate of cooling, the temperature range, the mixture of sperm and media, and the vials used. Presently, there is ample evidence that final storing of sperm is best performed in liquid nitrogen. This will provide a stable low temperature without fluctuations. After the discovery in 1953 by Bunge and Sherman that glycerol treated frozen and thawed human sperm could retain its fertilizing capacity, glycerol has been the essential additive to various protective media. From recent reports it has been claimed that the final concentration of it is optimal at 7.5%. It probably stabilizes the hydrated intracellular protein structure through its electrolyte- and water-binding capacity. In our laboratory we have chosen to modify the method of Behrman and Sawada (1966). The media contains egg yolk, glycerol, glucose, sodium citrate and glycine. For details the reader is referred to Dr. Fribergs paper in this issue.

Most workers dealing with freezing of human sperm have utilized glass ampoules for storage in liquid nitrogen. There are, however, some drawbacks to the use of glass ampoules. During the thawing procedure there is a great risk that the ampoules can burst, and the surface of the sample is comparatively small. This may, at least theoretically, result in heat dispersion to the central part of the sample when the crystallisation heat is liberated during the freezing procedure (-5 to -15°C). In order to increase the freezing area and thus to reduce this risk of heat dispersion to the central part with resulting sperm damage we have chosen to use "paillettes" a type of plastic straw (Instruments de Médicine Vétérinaire SA, L'Aigle, France). They

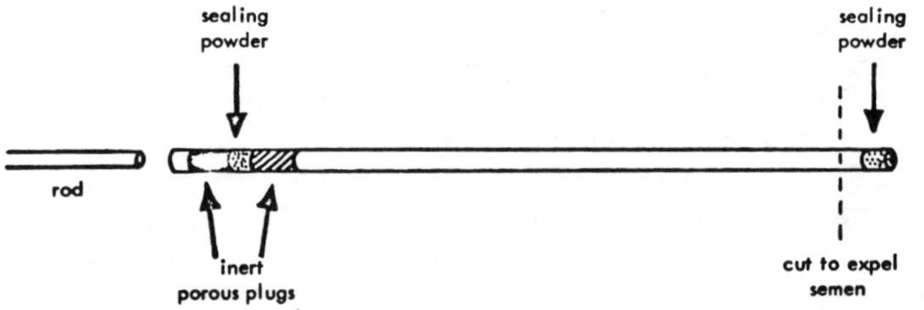

Fig. 1. Plastic straws used for freezing and storage of semen-media mixture. The rod is used to expel the sample after cutting the end from the tube.

(from Matheson et al. 1969)

are 13 cm long and contain 1/2 ml (Fig. 1). The straws have been widely used in veterinary medicine and have numerous advantages: They are easily sealed with plastic, are flexible, are easy to mark and store, and they can be cut easily with scissors. Dr. Friberg in his paper in this issue will give pertinant data on the freezing procedure and the rate of cooling.

For each insemination we have used the contents of one plastic straw, i.e. 1/2 ml. The sperm was deposited intracervically once daily. No intrauterine inseminations have been done, as this method not infrequently causes infection, uterine pain or both.

When using frozen semen for insemination proper timing and repeated insemination procedures are mandatory. Comparison between the in vivo survival time in the cervical mucus between fresh and frozen sperm has revealed that the survival time for frozen sperm is approximately 24 hours. The corresponding survival time for fresh sperm will normally be at least twice as long. The most frequently used criteria for the determination of ovulation are the temperature readings and the macroscopic features of the cervical mucus. It will be emphasized that the temperature readings have a limited value in the precise guidance for timing of ovulation. In doubtful cases it will be of great value to add direct hormone determinations. Fig. 2 depicts one woman of proven fertility who was investigated during one menstrual cycle. The cervical mucus was investigated for quantitative sperm penetration and sialic acid content. The day of probable ovulation was determined by the method of Johansson & Wide (1969) measuring the plasma progesterone concentration. The daily total urinary oestrogen excretion was also determined. The day of highest sperm penetration coincided with the midcyclical oestrogen peak and preceeded the day of probable ovulation by one day. On the day of ovulation the sperm penetration had decreased considerably from the peak value (Carlborg et al. 1969). This observation suggests that insemination should be performed just before ovulation. This also appears reasonable in the light of accumulating knowledge of the time necessary for capacitation of human sperm.

During a 3 year period we have treated a total at 76 women for 225 cycles. Twentytwo women became pregnant, 4 of which conceived twice. There were 26 pregnancies. Fig. 3 shows the distribution of treated cycles when conception occurred. It is remarkable that so many conceived during the first cycle. However, a similar distribution of data has been obtained by other workers regardless of whether frozen or fresh sperm was used. The important difference is that the pregnancy rate after the use of frozen sperm approximates 30-50% of the women treated within 12 cycles. The comparative figure for fresh sperm approximates 80%.

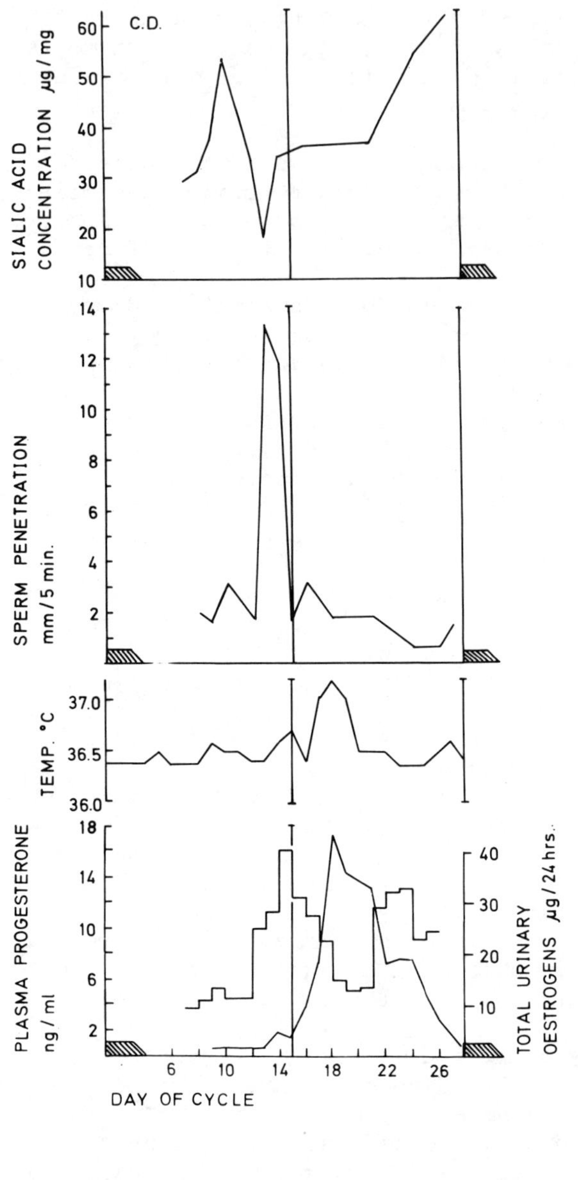

Fig. 2. Sperm penetration and sialic acid content of cervical mucus, basal body temperature, total urinary oestrogen excretion and plasma progesterone concentration during one cycle of a fertile woman. The day of probable ovulation is indicated by a vertical line.

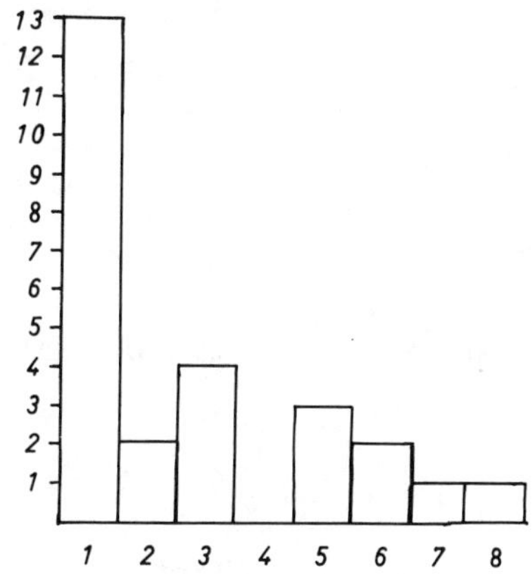

Fig. 3. Distribution of treated cycles needed until pregnancy occurred after insemination. On the abscissa number of cycles, on the ordinate number of treated women.

There is thus a definitely lower conception rate compared to that of fresh sperm when frozen sperm is used. For practical reasons the use of frozen sperm has numerous advantages. Further research into the improvement of freezing techniques and basic knowledge of sperm and reproductive physiology is needed before the increasing demand of both heterologous and homologous insemination can be met.

References:

Behrman, S.J. & Sawada, Y.: Fertil. & Steril. 17 (1966) 457.
Bunge, R.G. & Sherman, J.K.: Nature 172 (1953) 767.
Carlborg, L., Johansson, E.D.B. & Gemzell, C.: Acta endocr. (Kbh) 62 (1969) 721.
Johansson, E.D.B. & Wide, L.: Acta endocr. (Kbh) 62 (1969) 82.
Matheson, G.W., Carlborg, L. & Gemzell, C.: Amer. J. Obstet. & Gynec. 104 (1969) 495.

OXYGEN CONSUMPTION OF HUMAN SPERMATOZOA AND SEMINAL PLASMA

Eliasson, Rune

Department of Physiology, Faculty of Medicine

Karolinska Institutet, S 104 01 Stockholm

In the present investigation the oxygen consumption of human spermatozoa in the seminal plasma as well as in a buffered salt solution has been determined by using a polarographic technique. The analyses have been performed on single ejaculates from about 80 men. Correction has been made for the oxygen consumption of the seminal plasma in all experiments on whole ejaculates.

MATERIAL AND METHODS

Semen was supplied by patients referred for infertility and/or prostato-vesiculitis and also by healthy volunteers. The specimens were obtained by masturbation (in most cases) or coitus interruptus and brought to the laboratory within one (max. two) hour. All subjects had delivered at least two specimens for careful routine analyses before the experiments began.

The semen was analysed for volume, sperm density, motility, vitality (supravital staining) and morphology (Eliasson 1970). The sample was then divided into two parts, one being centrifuged to remove the spermatozoa (2800 g, 20 min.), the other being used untreated. The oxygen consumption was determined by polarographic measurements of the oxygen tension ("53 Biological Oxygen Monitor System", Yellow Springs) simultaneously in the seminal plasma and the whole semen. After the experiment (15 min.) new determinations of sperm density, motility, and vitality were carried out. The values given for sperm density, motility, and vitality are means of the two determinations.

The oxygen consumption was calculated as $\mu l/10^8$ spermatozoa per hour at 37°C and 760 mm Hg (Z_{O_2} spermatozoa) and $\mu l/ml$ per hour (Z_{O_2} seminal plasma) respectively, assuming that the oxygen content of the seminal plasma (and the salt solution) was 0.03 $\mu l/ml/mm$ Hg.

When washed spermatozoa were used the semen (1 vol) was diluted with a buffered Ringer solution (3 vol) and centrifuged at g for 20 minutes. The supernatant was aspirated and 3 ml of the salt solution added. After gentle mixing the sample was again centrifuged, the supernatant removed and 1 ml of buffered Ringer solution added.

In this study 20 men with a careful clinical examination including prostate massage and cytological examination of the prostate fluid are included. They have been classified into the following two groups.

A. Clinically healthy men (N = 12) - no history of recent illness and a normal prostate gland.

B. Prostato-vesiculitis (N = 8) - the rectal examination revealed a clearly pathological prostate and/or the cytological examination of the prostate secretion (after Papanicolaou staining) showed clear signs of inflammation.

The significance of intra-individual differences has been calculated by using student t-test for pair analysis and for inter-individual differences student t-test for unpaired data (Richterich 1969).

RESULTS

The Z_{O_2} for the spermatozoa and the seminal plasma from 82 samples (40 men) are summarized in Table I. The values are given for total number of cells, and after correction for non-motile and dead cells. The mean values and ranges for total sperm Z_{O_2} and seminal plasma Z_{O_2} correspond well with reports from others. There are no data available in the literature for Z_{O_2} after correction for dead cells. The marked variation in Z_{O_2} for the spermatozoa is due to inter-individual differences and not to intra-individual. The first and second samples from the same subject have, on the contrary, small differences, even if exceptions were noted.

	Z_{O_2} Spermatozoa			Z_{O_2} Seminal Plasma
	total cells	motile cells	living cells	
Mean	1.7	5.6	3.7	4.9
S.D.	1.62	7.83	4.05	1.89
Range	0-6.3	0.26.9	0-15.0	1.7 - 8.5
N	82	82	82	79

Table I. Oxygen consumption of human spermatozoa and seminal plasma in single ejaculates from 40 men. Z_{O_2} denotes μl $O_2/10^8$ spermatozoa/hour or μl O_2/ml seminal plasma/hour. Values are corrected for 37°C and 760 mm Hg.

The Z_{O_2} for the spermatozoa decreased with time when calculated from the total number of cells, but was constant, when corrected for the percentage of dead cells. It thus appears that oxygen consumption per 10^8 vital spermatozoa gives a more accurate estimate of the sperm respiration. There was a significant difference between the mean oxygen consumption of the spermatozoa in samples from healthy men (group A) and from men with prostato-vesiculitis (B) ($p < 0.01$); the Z_{O_2} per 10^8 vital cells being in the first group 4.3 and in the second one 1.4. Five out of eight patients with prostato-vesiculitis (group B) delivered semen samples with no (or "negative") oxygen uptake for the spermatozoa. If the "negative" values had been used, the difference between groups A and B had been more accentuated. Three patients with prostato-vesiculitis have after intensive treatment of their infections given new semen specimens in which the sperm respiration was normal.

To study if the depressed oxygen consumption in the specimens from patients with prostato-vesiculitis was due to changes in the spermatozoa or in the seminal plasma cells were washed and suspended in a buffered Ringer solution and the oxygen consumption determined. Five specimens from four patients, whos spermatozoa in the whole semen had been zero were used. In the Ringer solution the spermatozoa from three of the patients revealed high Z_{O_2} ($\bar{x}$ = 66) and those from the fourth patient had "normal" respiratory activity ($\bar{x} = 4.7$).

In another series of experiments it was studied if the increase in Z_{O_2} when the spermatozoa were transferred to a salt solution was a general phenomenon or not. The mean oxygen consumption per 10^8 live spermatozoa/hour increased from about 5 (whole semen) to 13 microliter (salt solution) ($p < 0.001$). If the cells were again suspended in the seminal plasma the oxygen consumption returned to the same level as before, indicating that the increase of Z_{O_2} could not be a result of the physical treatment of the spermatozoa.

DISCUSSION

So far Nevo (1966) is the only one, who has reported the respiratory rate of untreated spermatozoa in single ejaculates and also used a correction for the seminal plasma and the non-motile cells. The values reported by him correspond closely to those found in the present investigation as well as to those reported by Murdoch and White (1968), Peterson and Freund (1968), and Karjalainen and Niemi (1969).

It was noted that Z_{O_2} for the total number of cells markedly decreased with time after ejaculation, while the values corrected for the per cent dead cells were encouragingly constant. It therefore appears that Z_{O_2} calculated for the number of "vital" cells gives a satisfactory estimate of the oxygen uptake by the spermatozoa.

Peterson and Freund (1968) noted that in some specimens the oxygen uptake (Warburg technique) was the same as or even higher in the sperm free seminal plasma than that of the whole semen. The same observation was made in the present study. This phenomenon occurred in specimens from five out of eight patients with clinical and/or laboratory signs of prostato-vesiculitis. None of the healthy men had a Z_{O_2} below 1.0 (for living cells).

The low oxygen consumption of the spermatozoa in specimens from men with prostato-vesiculitis (and possibly other disturbances of the genital tract) is of more general interest. The phenomenon seems to be due to a factor(s) in the seminal plasma, since the washed spermatozoa from these specimens showed a high or normal Z_{O_2} when suspended in a buffered Ringer solution.

On the other hand, it was noted to be a general phenomenon that the oxidative metabolism of the spermatozoa increased when the cells were transferred from the seminal plasma to a buffered salt solution. The reason for this previously not reported change in the metabolism is unknown, but warrents further investigations.

SUMMARY

The oxygen consumption of human spermatozoa has been determined polarographically both in their natural environment and in a buffered Ringer solution. The mean oxygen consumption of the spermatozoa ($\mu l\ O_2/10^8$ live cells/hour) was in native single samples from healthy men 4.3 (SD $\pm$ 3.4) and from men with prostato-vesiculitis 1.4 (SD $\pm$ 2.5). The difference was significant ($0.01 > p > 0.005$), and seems to be due to an inhibitory factor(s) in the seminal plasma.

The oxygen consumption of washed spermatozoa was two to three times higher in a buffered Ringer solution than in the whole semen ($p < 0.001$). The reason for this previously not reported difference is not known.

ACKNOWLEDGEMENT

Financial support has been received from the Swedish Medical Research Council (14x-538) and the Population Council. The skilful technical assistance of Miss A. Björklund is also thankfully acknowledged.

REFERENCES

Eliasson, R. (1970) Standards for investigation of human semen. Andrologie (to be published)

Karjalainen, K., and Niemi, M. (1969) Oxygen consumption of human spermatozoa and seminal plasma as measured by a micro-diver technique. Scand. J. clin. Lab. Invest. 23, suppl. 108, 79.

Murdoch, R.N., and White, I.G. (1968) Studies of the metabolism of human spermatozoa. J. Reprod. Fertil. 16, 351-361.

Nevo, A.C. (1966) Relation between motility and respiration in human spermatozoa. J. Reprod. Fertil. 11, 19-26.

Peterson, R.N., and Freund, M. (1968) An evaluation of the respiratory capacity of human spermatozoa. J. Reprod. Fertil. 19, 351-372.

Richterich, R. (1969) Clinical Chemistry. Theory and Practice. Ed. S. Karger, Basel. Academic Press, New York.

EXTRACELLULAR GLUTAMIC-OXALOACETIC TRANSAMINASE AS A MEASURE ON MEMBRANE INJURY IN SPERMATOZOA DURING TREATMENT

B.G.Crabo, R.E.Bower, K.I.Brown, E.F.Graham & M.M.Pace

Department of Clinical Biochemistry, Roy. Veterinary College, 10405 Stockholm, Sweden and Department of Animal Science, Univ. of Minnesota, St.Paul, Minn.55101, U.S.A.

Since bull semen was first frozen by Polge et al. (1949) sperm motility has been the major criterion for evaluation of the effectiveness of the freezing process. For bull semen motility is generally considered a fairly reliable indication of the fertilizing capabilities of fresh or frozen semen. However the reliability of motility as an index of semen fertility is swiftly eroded when we move to other economically important species, particularly swine and turkeys. Efforts to develop frozen semen for these two species have been confounded by good sperm motility after freezing, but practically no offspring have resulted from insemination.

Unquestionably the best test of freezing techniques is fertility trials, but the high costs in both time and money render them economically unfeasible. This is particularly true when the starting point is zero fertility. The need for an inexpensive, quick method to evaluate semen preservation techniques is therefore apparent.

Fertilization involves processes that are intrinsically related to sperm cell membranes (e.g. sperm capacitation and the liberation of acrosomal enzymes). The integrity of the sperm cell membrane would therefore be important in successful fertilization and any trauma to the membranes, such as disruption by ice crystal formation, would detract from their ability to function normally.

In our search for a new method of semen evaluation we have looked at concentration changes of intracellular substances in the extracellular media as a function of sperm cell membrane integrity. Analyses of seminal plasma for glutamic-oxaloacetic transaminase (GOT) have indicated the extracellular concentration of this enzyme might serve well as an index of membrane trauma, since it occurs abundantly in the semen of the species under study and is relatively easy to determine (cf. Graham and Pace 1967; Brown et al., 1970a). While GOT does not render information concerning the integrity of the acrosome that hyaluronidase might, the determination of GOT is easier and more reliable. In addition, other experiments have indicated changes in extracellular hyaluronidase parallel those of the transaminase (unpublished observations).

Methodology

Pace and Graham (1970) showed that centrifugation alone could cause an increase in extracellular GOT. Since it was desirable to separate the spermatozoa from the seminal plasma without causing an additional release of GOT, Brown et al. (1970) examined various separatory methods and found a density gradient centrifugation which permitted a minimal GOT release. The aqueous gradient fluid containing 15% raffinose and 6% dextran (M. W. 15-20,000) was isotonic and had a specific gravity between that of seminal plasma and spermatozoa. If a cryophylactic agent such as glycerol or ethylene glycol had been added to the diluted semen, then it was also added to the density gradient at the same concentration to maintain the tonicity and correspondingly increase the specific gravity. The gradient fluid was placed in tubes and the neat or diluted semen was carefully pipetted on top under isothermal conditions. Under these conditions there was minimal thermal and osmotic damage to spermatozoa at the interface. The tube was then centrifuged at 12,000g for 10 minutes and the supernatant was carefully removed and analyzed for GOT. The results were expressed in International Units of GOT per 10^9 spermatozoa. The maximum releasable GOT was defined by directly plunging semen samples into liquid nitrogen.

Effects of buffers on the release of GOT

The effects of some buffers, e.g. citrate, phosphate and Good's buffers: Tricine and TES (Good et al. 1966), on sperm motility and GOT release in boar semen illustrated the usefulness of the technique.

The effects of the four buffers illustrated in Fig. 1 are typical of 30 combinations examined (Graham et al. 1970; Crabo

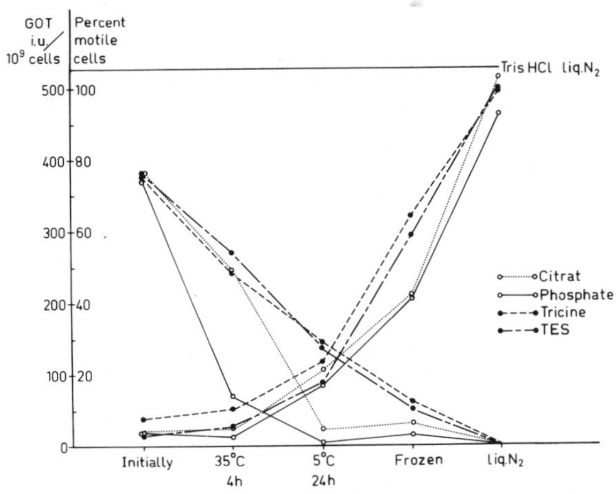

Fig. 1. Average motility and extracellular GOT after various treatments of boar semen in four diluents, all containing 20% egg yolk and 0.2 % fructose. There were statistically significant differences ($p < 0.05$) between all buffers except Tricine and TES for motility.

et al., 1970; and Brown et al., 1970). Each buffer contained 0.2 per cent fructose and 20 per cent fresh egg yolk. Motility and GOT release were measured after the initial dilution, after 4 hours at 35°C, after 24 hours at 5°C, after the addition of glycerol (6% v/v) and pellet freezing on dry ice, and after directly plunging samples into liquid nitrogen.

While there appears to be a negative correlation between motility and GOT release over all buffers this relationship was not statistically significant within individual treatments except for samples stored at 5°C. It is possible that motility changes represented the treatment effect on the middle piece of the spermatozoa while the extracellular GOT, represented its effect on the sperm cell membrane. Further evidence supporting this concept of dual effects was demonstrated by the results of treatments on sperm diluted with citrate and phosphate buffers. GOT release was significantly lower in these two buffers than in any other but sperm motility was significantly lower at 35°C in phosphate and at 5°C motility was significantly lower in both phosphate and citrate buffers. On the other hand spermatozoa in TES and Tricine extenders maintained greater motility ($p < 0.05$) during incubation, cooling and freezing but demonstrated a significant increase in extracellular GOT after freezing. Tricine in particular showed the highest GOT release for the four treatments, but maintained better motility than citrate and phosphate buffers.

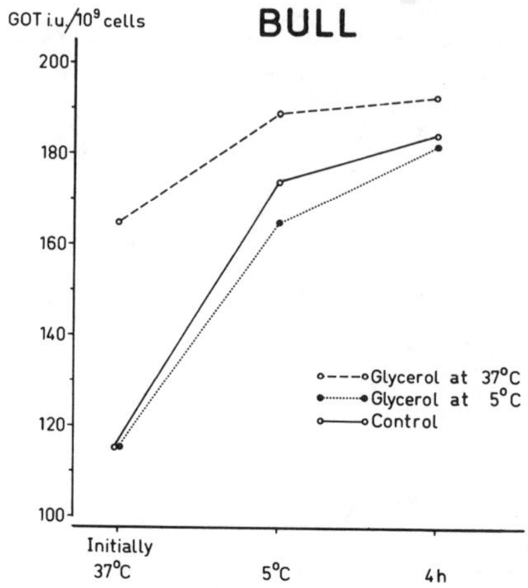

Fig. 2. Average extracellular GOT in bull semen extended 1:15 with Tricine-Tris buffer containing 20% egg yolk and 2.5% fructose after addition of 5% glycerol initially (37°C) or after cooling to 5°C.

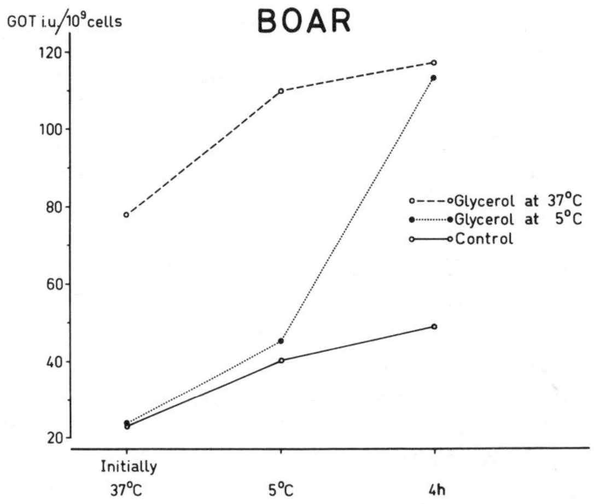

Fig. 3. Average extracellular GOT in boar semen (neat, extended 1:1 and 1:4 in Tricine-Tris buffer) with 20% egg yolk and 2.5% fructose) after addition of 5% glycerol initially (37°C) or after cooling to 5°C.

The difference between the effects of TES and Tricine buffers and citrate and phosphate buffers was believed to be due to the difference in ionic strength (0.07 and 0.03 for the former and 0.24 and 0.36 for the latter) or the affinity for metal binding of citrate and phosphate (cf. Good et al. 1966).

Comparisons with neat semen showed that dilution in any buffer caused a release of GOT into the media. Tricine caused a slightly higher release than the others tested. This initial loss of GOT was thought to be associated with the dilution effect cited by Mann (1964).

Effect of glycerol on the release of GOT

Bull semen in preparation for freezing can be diluted directly in a glycerated diluent or the glycerol can be added dropwise in the coldroom to the diluted semen without affecting its fertilizing ability (Graham et al., 1958; The Milk Marketing Board, 1968). Likewise, glycerol does not appear to have an adverse effect on human semen (cf. Sherman, 1954). However, when glycerol was added to boar semen, fertility was depressed markedly even without freezing the semen (Polge, 1956; King & MacPhearson, 1966) while the effects on motility were minimal (Polge, 1956; Hoffman, 1959). To clarify this difference between bull and boar semen a trial was conducted.

Extracellular GOT was examined to study the contrary effects of glycerol on bull and boar semen. Both neat semen (boar only) and semen diluted in a Tris-Tricine buffer (3.5% Tris, 50-66% Tricine, 2.5% fructose 20% fresh egg yolk) were used. GOT was measured initially at 37°C, after cooling to 5°C, and at the end of four hours equilibration at 5°C. Glycerol was added (5% v/v) to samples initially and after cooling to 5°C. The controls contained no glycerol.

Figure 2 illustrates the effects of glycerol on bull semen in terms of GOT release. There was an increase in extracellular GOT with time and with cooling. Glycerization at 37°C caused a significant ($P < 0.05$) increase in GOT which remained significantly greater than those glycerated after cooling to 5°C. At the end of the equilibration time no difference could be shown between any of the samples.

The results of the corresponding experiment conducted with boar semen are illustrated in Fig. 3. As in bull semen, samples glycerated at 37°C demonstrated a significant increase in extracellular GOT. Addition of glycerol after cooling caused a slight but non-significant increase in GOT but at the end of four hours the GOT levels had essentially eqaled those of samples glycerolated initially and all samples containing glycerol showed a significant increase in extracellular GOT over the controls. This is in marked contrast to what was seen in bull semen.

These findings are in agreement with results obtained from breeding trials (cf. above) and thus suggest parallels between extracellular GOT and semen fertility. Further studies on the effects of glycerol on bull and boar spermatozoa (Bower et al., 1970) have indicated that boar spermatozoa appear to be more sensitive to osmotic changes than those of the bull.

References

Bower, R.E., Crabo, B.G., Pace, M.M. & Graham, E.F.: Effect of glycerol on the release of glutamic-oxaloacetic transaminase from bull and boar spermatozoa. To be published 1970.

Brown, K.I., Crabo, B.G., Graham, E.F. & Pace, M.M.: Some factors affecting loss of intracellular enzyme from bull, boar and turkey spermatozoa. To be published 1970.

Brown, K.I., Graham, E.F. & Crabo, B.G.: Effect of hydrogen ion buffers on storage and freezing of spermatozoa. II. Turkey. To be published 1970.

Crabo, B.G., Brown, K.I. & Graham, E.F.: Effect of hydrogen ion buffers on storage and freezing of spermatozoa. III. Boar. To be published 1970.

Good, N.E., Winget, G.D., Winter,W., Connolly, T.N., Izawa, S. & Singh, R.M.M.: Hydrogen ion buffers for biological research. Biochemistry 5, 467-477, 1966.

Graham, E.F. & Pace, M.M.: Some biochemical changes in spermatozoa due to freezing. Cryobiology 4, 75-84, 1967.

Graham, E.F., Crabo, B.G. & Brown, K.I.: Effect of hydrogen ion buffers on storage and freezing of spermatozoa. I. Bull. To be published 1970.

Graham, E.F., Voigt, D.W. & Fisher, G.R.: Effect of method of glycerol addition on the fertility of frozen bovine spermatozoa. J. Dairy Sci. 41, 1553, 1958.

Hoffmann, H.H.: Konservierung von Ebersperma durch Tiefgefrierung. Inaug.Diss. München 1959.

King, G.J. & Macphearson, J.W.: The effects of glycerol on fertility of liquid boar semen. A.I. Digest 14, 94-95, 1966.

Mann, T. Biochemistry of the semen and of the male reproductive tract. Metheuen, London, 1964.

Pace, M.M. & Graham, E.F.: To be published 1970.

Polge, C.: Artificial insemination in pigs. Vet.Rec. 68, 62-76, 1956.

Polge, C., Smith, A.U. & Parkes, A.S.: Revival of spermatozoa after vitrification and dehydration at low temperatures.

Sherman, J.K.: Freezing and freeze-drying of human spermatozoa. Fertil.Steril. 5, 357-371, 1954.

The Milk Marketing Board (England): Annual Report 1968.

THE EFFECT OF TUBAL SPERM CONCENTRATION ON THE RATE OF DEVELOPMENT OF SWINE OVA AND BLASTOCYSTS

Alanko,Matti

College of Veterinary Medicine, Depart. of Obstetrics and Gynecology

Hautjärvi,Finland

The number of spermatozoa in the uterine tubes of the sow varies between 10^1-10^4, debending on the amount of semen deposited and the efficiency of the sperm transport (Alanko,1968).
The fertilization of ova starts about 2-6 hrs.,and is completed at the latest 10 hrs. after insemination, when at least 90-95% of the ova have been fertilized (Hancock,1962).If the sperm concentration (SpC) is low in the uterine tubes (number of spermatozoa below 5 x 10^2),the fertilization percentage will be considerably lower, while the number of abnormally developing ova will increase (Alanko,1968).
Normally,the ova will be at least at the 4-cell stage on reaching the uterus but ova at only the 2-celled stage have been found in the uterus after delayed inseminations (Hunter 1967) or after low SpC in the tubes (Alanko 1968).
An investigation on the effect of tubal sperm concentration on the rate of develpment of fertilized swine ova was carried out at Hautjärvi during the years 1965-68.The material consisted of 54 gilts which were inseminated in various ways and of which 43 were slaughtered 1-3 days and eleven 8-11 days after insemination.All the gilts were inseminated about 24 hrs. after the beginning of the heat (=before the ovulation)

Results. 1. The location of ova and blastocysts. The first ova reached the uterus about 36 hrs. after ovulation. No ova were found in the tubes after 63 hrs After entering the uterus, the eggs were located in the apices of the uterine horns for some days; from the 8th day onwards, the eggs started to spread in the two uterine horns. The migration of the blastocysts from one horn to the other could be observed at the earliest on the 9th day (1 blastocyst out of 13). This phenomenon was most common on the 10-11th day (7 out of 50 and 12 out of 49, resp.). The migration from the left to the right horn was more common as the ovulation frequency is greater in the left ovary. On the 11th day, the blastocysts were evenly distributed in the horns in spite of the differences in the ovulation frequency (fig.1).

Fig.1. The distribution of blastocysts in the uterine horns.

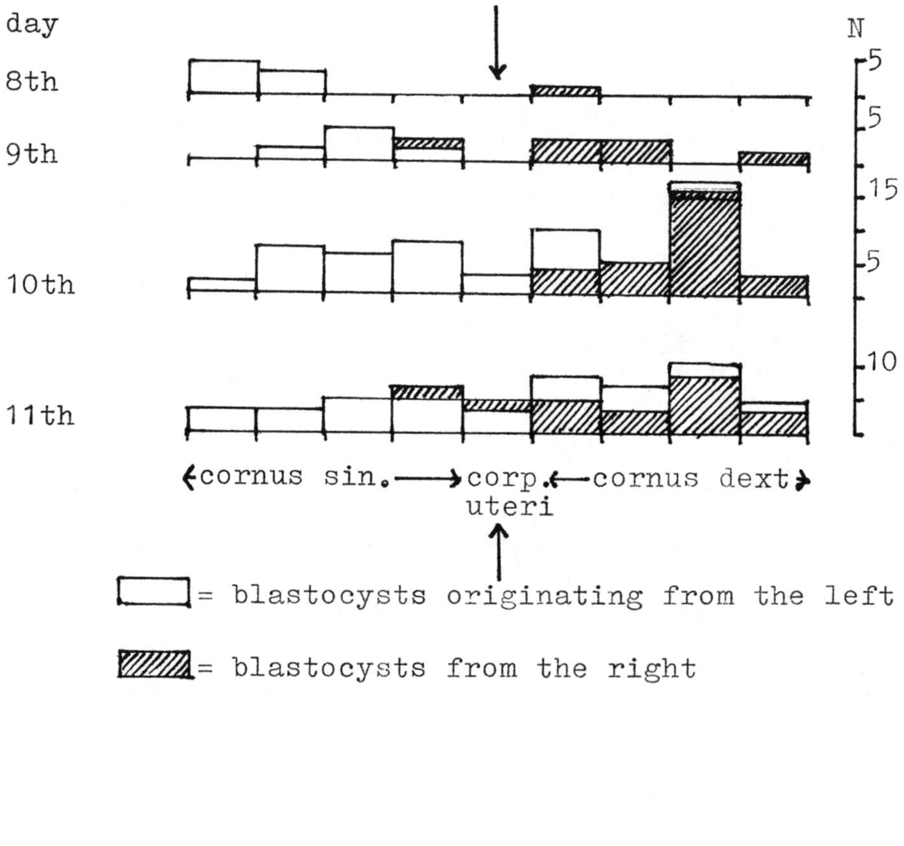

2. The development of fertilized ova in gilts slaughter- 1-3 days after the insemination is seen in fig. 2. On an average, the ova in tubes with high SpC were at a more advanced stage of development thanthose in tubes with a low number of spermatozoa ($p<0,001$).

Fig. 2. The appearance of ova at different stages of development in relation to time elapsed since ovulation

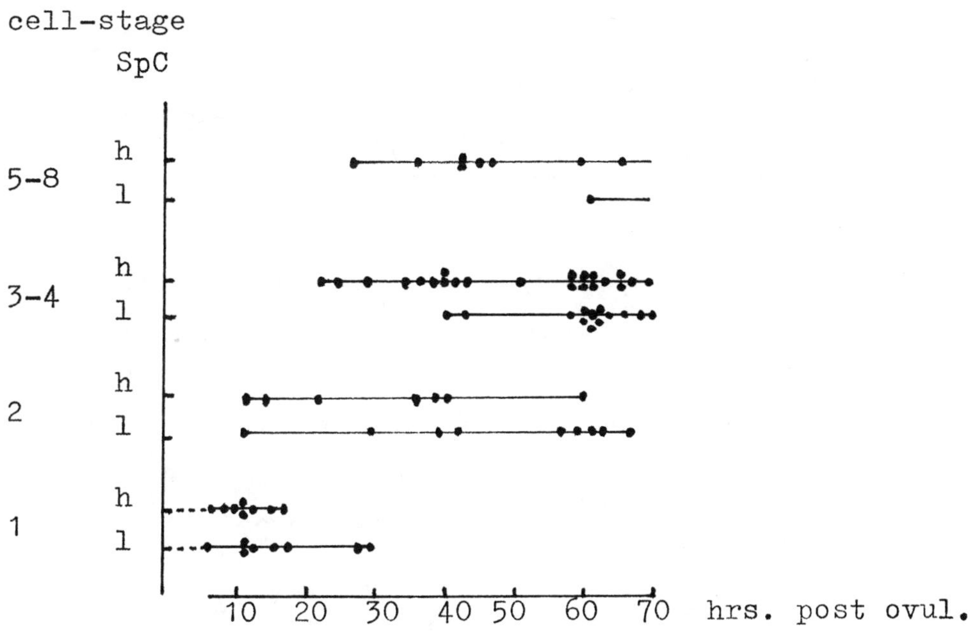

The difference in development between ova from tubes with high or low tubal SpC tended to grow with advanced stage of development.This difference was 13 hrs. between the last 1-celled ova, 18 hrs. between the first 3-4-celled ova and already 35 hrs. between the first 5-8-celled ova.
The slower cleavage-rate in connection with the low tubal SpC is also evident in ova found in the uterus (table I).
The sizes of blastocysts between the 8th and the 11th day are illustrated in fig. 3 and shown in table II. On account of the limited material,no statistically significant difference in development between blastocysts originating from tubes with high or low tubal sperm concentration can be shown but the trend was

about the same as that observed in the case of 1-3 days old ova.

Table I. The cleavage-stages of ova found in the uterus 59-70 hrs. after ovulation (20 gilts, 110 ova).

SpC in tubes	cell-stage of ova				total
	2	3	4	5-8	
high	3	5	50	20	78
low	13	8	10	1	32
	16	13	60	21	110

$p < 0{,}001$

Fig 3. The size of blastocysts at different ages.

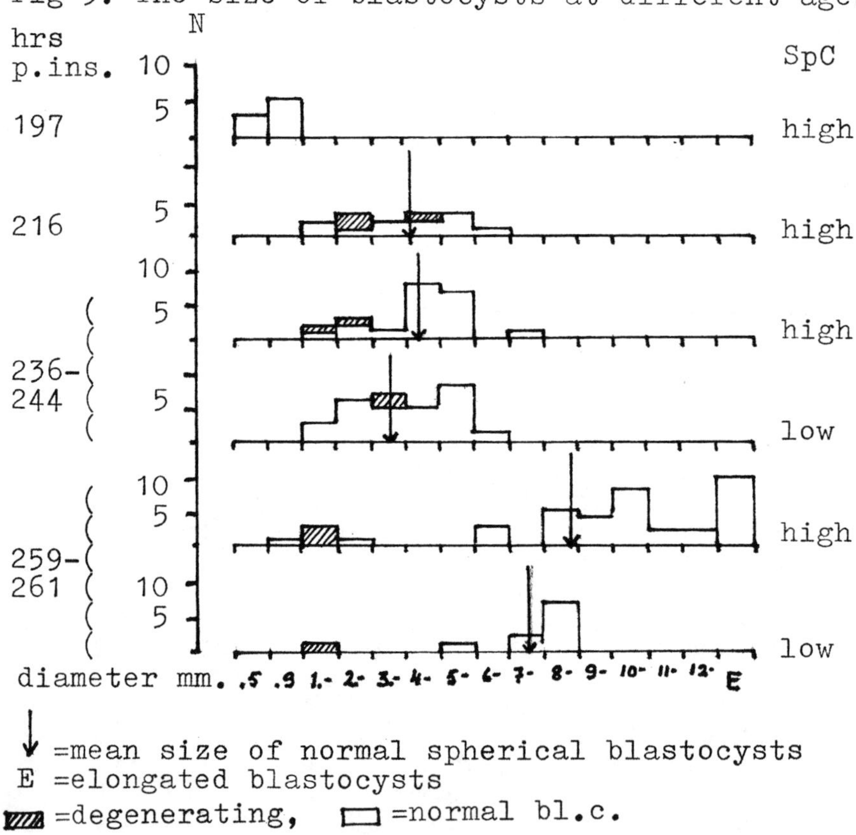

Table II. The size of blastocysts at different times after insemination.

time hrs.	SpC	N gilts	N c.l.	N bl.c.	spherical bl.c. normal	mean size mm.	degen.	elongated bl.c.
197	high	1	11	8	8	< 1	-	-
216	high	1	14	14	9	4,1±2,2	5	-
236-244	high	3	37	22	20	4,5±1,3	2	-
	low	2	32	30	28	3,7±1,4	2	-
259-261	high	3	42	39	26	8,8±2,4	3	10
	low	1	15	11	10	7,7±0,9	1	-
		11	151	124	101		13	10

Discussion. It is known in AI-practice that the proportion of small litters is greater in artificially inseminated gilts than in naturally served ones. It has also been observed that the spermtransport may fail as a result of diverse disturbing factors, which are more frequent in connection with A.I. The failure of spermtransport results in a low SpC in the tubes, which lessens the possibility of the ova to encounter the sperm, thus prolonging the time of fertilization.

For many mammalian species (mouse, rat, rabbit), it has been established that embryonic survival depends upon the egg reaching a critical stage of development at the time the uterus is ready receive it. Presumably, an appreciable degree of embryonic mortality may be accounted for by the lack of synchrony between ovular and uterine development (Gates, 1965).

In the present study, the retardation of development was much greater in ova from tubes with low SpC than in ova from tubes with high SpC. It seems obvious that the retardation in development of ova from tubes with low SpC may be accounted for to a high degree by the delay in time of penetration of the egg by a spermatozoon. This delay could be sufficient to put the development of the egg out of synchrony with that of the uterine endometrium, a condition previously known to lead to embryonic

mortality (Smidt et al.1964).
The comparatively often observed low SpC in the tube in connection with artificial insemination thus seems to cause a decrease in litter size in three ways:
1) by lowering the proportion of ova fertilized,
2) by increasing the share of abnormally developing ova,
3) by causing a delay in the time of fertilization of the egg with a subsequent retarded cleavage rate and asynchrony between ovular and uterine development.
In the two last mentioned cases the emryonic mortality during the early pregnancy is increased.

References:

Alanko,M and Moberg,R (1968) Fortpfl.Bes.Aufzucht der Haustiere 4:1,16-20
Alanko,M (1968) VIe Cong.Int.Reprod.Anim.Ins.Artif. Paris, vol. I, 615-617
Gates,A (1965) Preimplantation Stages of Pregnancy, Ciba Foundation Symposium,Churchill, 270-288
Hancock,J.L. (1962) Anim.Breed.Abstr. 30,3 pp.285-310
Hunter,R.H.F. (1967) J.Reprod.Fert. 13,133
Smidt,D and Steinbach,J. (1964) 5th Int.Congr.Anim. Reprod.Art.Ins. VII, 527 (Trento)

THE PRACTICAL USE OF DF-SEMEN ON CATTLE IN SWEDEN

Thure Swensson

Uppsala läns Seminförening,Enköping,Sweden

The reason for the use of AI on cattle is still the same as it was in the beginning of the forties,namely a lower cost of insemination by good quality bulls,especially on small farms; the possibility of reducing the incidence of venereal diseases; and finally genetic improvement of livestock by the intensive use of progeny tested bulls,i.e.bulls of known genetic value.With regard to the latter,Polge et al.(1) stimulated much interest when they published their first work on deep-frozen semen.It would be possible to make planned matings independent of time and distance and would also be possible to carry out breeding polices on a larger scale.

The English methods were introduced in Sweden in 1952 and the bull studs began using deep-frozen semen. As in other countries the results were on average 5-10 % lower than for fresh semen and therefore we used deep-frozen semen only as a complement to fresh semen for about twelve years. Now we know something of the errors we made. For example the choice of ampoules used had considerable influence on the results.Only two types of ampoules were proof against breakage and against leakage of alcohol from the freeze-bath.

When Nagase and Niwa(2) showed that it was possible to get better results with freezing rapid in small doses and when Jondet (3) and Cassou (4) had developed the plastic straws,there was new hope in introducing a DF-program on a full scale. Field trials in this country were carried out by Ström (5) with pelleted semen and Dyrendahl (6) with straws during 1964 and 1965.The fertility results were similar and very promising when compared to those of fresh semen.

The bull studs in Sweden chiefly use the French straw for 3 reasons:
a/ identification of each single dose was possible by printing,
b/ the semen was hygienically covered by plastic material,
c/ a special instrument could be used for the inseminations, which facilitated the field work. Since dec. 1967 all bull studs except one use straws while 95 % of the inseminated cows are inseminated with straws and 5 % with pelleted semen. About 600 000 cows are inseminated yearly.

Routine Work.

Before young bull calves are put into use, they are tested for identity by bloodgroup, for health (including blood test for chromosome anomalies), and later for serving ability and semen quality. Also the freezing properties are tested and finally a field test of fertility with 300 inseminations is performed. At about 15 months of age when proved the bulls are used as much as possible: once a week or every 5th day depending on capacity. Two ejaculates are collected within 5-10 minutes each time the bull is used.

The semen is assessed for volume, motility, cell content etc. and concentration in an absorbtion photometer. It is then diluted in Laiciphos (consisting of milk, sugar, antibiotics) and eggyolk to get a final dilution of 20-30 millions per dose. It is cooled for 1 hour (or even less) to + 4°C. The glycerol fraction of the diluter is then added to the solution drop by drop during a half hour period. The final glycerol concentration is 5 %. 5 hours are necessary for equilibration and during this time the straws are filled up and sealed with alcoholic polyvinylchloride-powder. The powder becomes firm on coming in contact with the semen and from the outside also in contact with water in a waterbath at + 4° C. The straws are then dried and freezed in liquid nitrogen vapour for 7 minutes and finally put into liquid nitrogen for storage.

It is possible to freze 8-10 000 straws a day in a relatively small laboratory when semen is collected from 10-15 bulls. Pruduction costs about 35 öre per straw and storage in big containers about 4 öre per straw per day.

The breeding program undertaken includes the freezing of 30 000 doses from each young bull, the immediate use of 1000 doses for progenytesting, storage of the rest until the progeny test is complete and then general use if the bull is approved. When 30 000 doses are frozen the bull is slaughtered.

AI with French straws is carried out with specially designed equipment. The farmer rings to the centre in the morning and the technicians visit the farms taking semen stored in thermosflasks containing liquid nitrogen. Thawing is performed 15 seconds in water at + 34° C.

Our AI-service also includes sterilitytreatment and pregnancy diagnosis. The staff includes veterinarians and technicians.

Recording of Fertility.

All AI:s made are registered in a central computer and the following information is then available for each bull:

% 28 days NR (non return)
% 56 " "
% 168 " "
% 273 " "

Moreover about 60 % of all cows inseminated are examined for pregnancy by rectal investigation.

Some field trials has been performed during the first years of use in this country. With deep-frozen semen it is possible to study seasonal variation in bulls' fertility apart from the influence of seasonal variation in cows' fertility. A slight variation in bulls fertility was found by Ström (7) while the highly significant seasonal variations were due to the cows.

The influence of storage time on fertility of pelleted semen in liquid nitrogen has been studied by Ström (8). No influence was found within one year of storage.

The straws now in use are of medium type (vol. 0.5 ml) but in order to reduce the space required for storage and to obtain more rapid freezing throughout the semen, efforts are being made to reduce the volume, i.e. to reduce the size of the straw. Malmberg and Dyrendahl (9) found no difference in fertility comparing medium-straws (0.5 ml) and mini-straws (0.25 ml).

The influence of equilibration times on fertility has been studied at the bull station at Enköping. (Table 1) Semen from 6 bulls was used in this trial and no difference could be found between deep-frozen semen equlibrated for 18 hours with 10 % glycerol and deep-frozen semen equlibrated for 5 hours with 5 % glycerol.

Table 1. Field trial. Two different equilibration times. Enköping 1969.

I: 18 hours equil. 10 % glycerol.
II: 5 " " 5 " "
No. of bulls: 6.

Time	AI	NR		Expected NR	
	No	No	%	No	%
I	2168	1288	59,4	1315.29	60.67
II	2935	1745	59.5	1751.56	59.68

However the experiment was a small one and must be confirmed.

Another small experiment performed at Enköping was the comparison between semen doses of 40×10^6 and 30×10^6 inseminated as deep-frozen semen in medium-straws. Three bulls were used and no significant difference in fertility results measured as 56 days NR-% was found. (Table 2)

Table 2.Field trial. Number of sperms. Enköping 1968.

Bull	40×10^6		30×10^6	
	AI No	% NR 56 d.	AI No	% NR 56 d.
85 R.	376	63.6	1064	63.1
1260 R.	465	66.9	291	70.1
30 W.	525	70.3	1819	66.6
Total	1366	67.3	3174	65.7

The figures indicate that the semen of different bulls may react in different ways when the number of sperm inseminated is changed. But this experiment also requires confirmation.

More research is required especially concerning diluters, glycerol content, equilibration times and also size of straws in order to facilitate production and to reduce storage costs.

References:

(1) Polge,C.,L.E.A.Rowson: Longterm storage of bull semen frozen at very low temperatures (-79° C) II Int.Congr.Anim.Reprod. 1952. Vol.III.90.

(2) Nagase,H.and T.Niwa: Studies on deepfreezing technique.III. Deepfreezing of bull semen in tablet form.1963.Jap.J.Anim. Reprod.,9,73.

(3) Jondet,R.:Congelation rapide du sperme de taureau conditionne en paillettes.1964.V.Int.Congr.Anim.Reprod.Vol.IV.463.
(4) Cassou,R.:La méthode de paillettes en plastique adaptée à la generalisation de la congelation.1964.V.Int.Congr.Anim.Reprod. Trento.Vol.IV,540.
(5) Ström,B.:Deepfreezing technique and fertility of bull semen frozen in pellet form. 1966. X.Nordic Veterinary Congr. Stockholm,Vol II,706.
(6) Dyrendahl,I.: The French straw -the payette- in frozen semen practice.1966. X.Nordic Veterinary Congr.,Stockholm.VolII,740
(7) Hultnäs,C.A.,B.Ström: Study on seasonal variation in bull fertility.1968.VI.Int.Congr.Anim.Reprod.Vol.I,281.
(8) Ström,B.: Influence upon fertility of bull semen of storage time in liqu.nitrogen. 1968.VI.Int.Congr.Anim.Reprod.Vol.II, 1171.
(9) Malmberg,G.,I.Dyrendahl: A comparison of the fertility of semen frozen in medium-and mini-straws.1969. Nordisk Veterinärmedicin,2,115.

A MALE SEXUAL CYCLE

J.E.Kihlström

Institute of Zoophysiology, Uppsala, Sweden

For a long time it has been looked upon as a fact that the female sexual functions vary cyclically, while the male sexual functions do not (Bullough 1951). This conception seems to be supported by recent studies of the sexual differentiation of the brain during the prenatal period of life (Harris 1964). However, the existence in both sexes of similar feedback mechanisms of interacting gonadotrophic and gonadal hormones makes it plausible to expect the occurrence of cyclic phenomena, corresponding to the oestrous cycle, also in the male. Such male sexual cycles, having in most species approximately the same length as that of the cycle of the corresponding females, have been demonstrated in rabbits (Doggett 1956, Kihlström 1958), cattle (Kihlström 1962), rats (Kihlström 1965), mice (Järnebrand and Kihlström 1969), sheep (Tischner 1967), and occur possibly also in man (Doggett and Keilers 1962, Exley and Corker 1965, 1966, Halberg et al. 1965, Hornstein 1965, Månsson 1965).

The most important parameters shown to vary cyclically are the volume of the ejaculate (Kihlström 1958, Degerman and Kihlström 1961, Kihlström 1963, Kihlström and Hultnäs 1970, Degerman and Kihlström 1970), the number of sperm cells (Doggett and Keilers 1962, Kihlström 1963, Kihlström and Hultnäs 1970), the sexual drive (Degerman and Kihlström 1961, Lisk 1969), the fertility (Kihlström 1962, 1963, Kihlström and Hultnäs 1970), the exfoliation of cells from the male urethra (Kihlström and Hornstein 1964 a, 1964 b, Hornstein 1965, Järnebrand and Kihlström 1969), the spontaneous ejaculations (Kihlström 1965), the body temperature (Degerman and Kihlström 1964), the secretion of seminal fructose (Degerman and Kihlström 1970), and the occurrence of palm-leaf-like crystals in the seminal fluid (Kihlström and

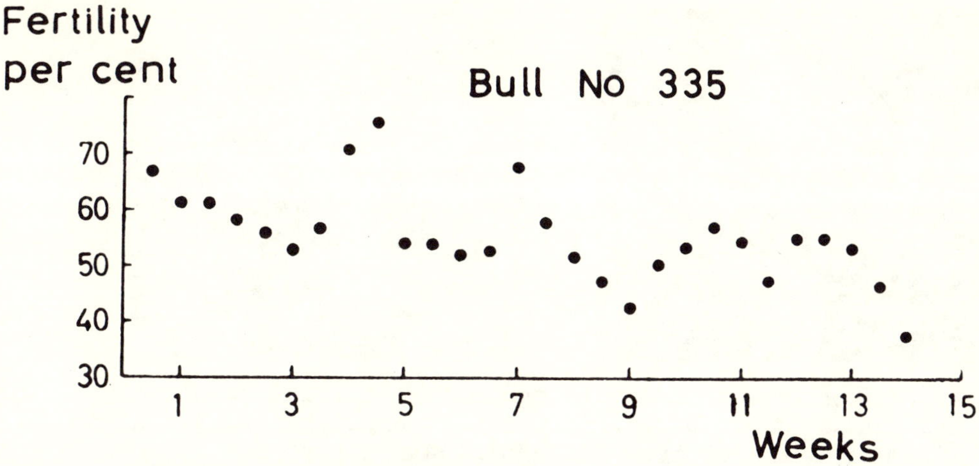

Fig.1. Variation of fertility with time in one bull. Fertility was defined as the number of observed pregnancies in per cent of the number of inseminations performed with each single ejaculate.

Fjellström 1969 a, 1969 b). Some experiments also indicate, that most of theses cyclic phenomena are different expressions of a corresponding fluctuation in the hormonal status, most likely in the circulating quantities of male sexual hormones (Hornstein et al. 1964, Exley and Corker 1965, 1966, Halberg et al. 1965, Inoué 1965 a, 1965 b, Degerman and Kihlström 1970).

When using semen for artificial insemination there are two important questions in this connection:
1. How does fertility vary during the male sexual cycle?
2. How to determine the phase of the cycle in order to have maximum chances of conception?
The first question has been studied in bulls used regularly for artificial insemination. One single ejaculate from such a bull may under optimal conditions be distributed among more than 100 cows. It is thus possible to get information about the fertilizing capacity of each single ejaculate.

The curve in Fig. 1 demonstrates the variation of fertility with time in one bull. Fertility was defined as the number of observed pregnancies in per cent of the number of inseminations performed with each single ejaculate. From this particular bull semen was collected regularly twice a week. As seen from the figure the length of the cycle is about 3 weeks. The mean difference between maxima and minima calculated from all bulls studied is 19 per cent, the highest value found in one bull being 43 per cent. There seems to be a difference in phase between this cyclic variation in fertility and the cycles observed in the

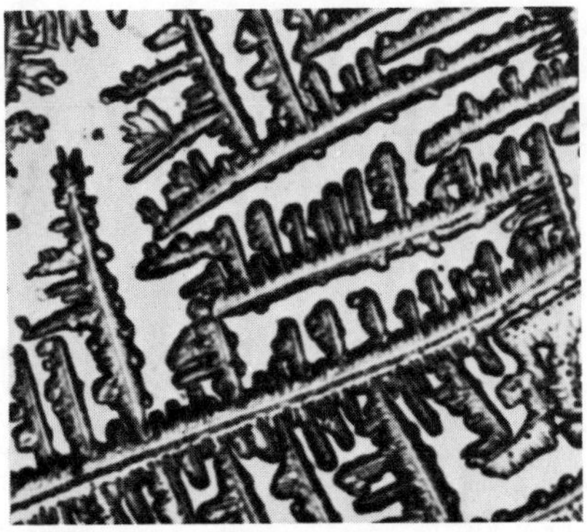

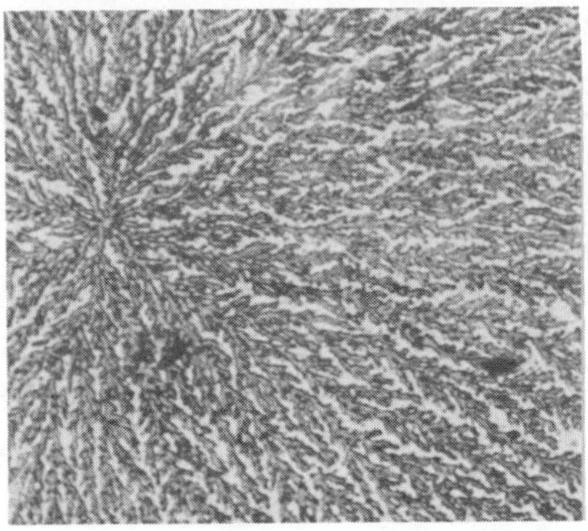

Fig.2. Micrographs of crystals occurring in dried samples of seminal fluid from rabbits. To the left crystals of type R, to the right crystals of type A. Magnification 1630 X.

volumes of the ejaculates and their amounts of sperm cells. Thus, peaks in seminal volume and number of sperm cells per ejaculate are followed some days later by a peak in fertility (Kihlström and Hultnäs 1970).

In order to have maximum chances of conception it might thus be of great importance to collect semen during the phase of the cycle having maximum fertility. This fact stresses the interest of the second question, viz. how to determine the phase of the cycle. Which one of the cyclically varying parameters mentioned before is most suitable for this purpose? Unfortunately some of them, e.g. the volume of the ejaculate and the number of spermatozoa vary with extraneous factors as well. On the other hand the study of e.g. the exfoliation of cells from the urethra is very time-consuming and the manipulation necessary might interfere with the sexual drive of the animals.

However, as mentioned before arboraceous crystals occur upon drying of seminal fluid (Kihlström and Fjellström 1969 a, 1969 b). These crystals are easy to prepare and easy to study. In such preparations of seminal fluid from rabbits two distinctly different types of crystals have been observed, the one called type R(right-angled crystals, Fig.2), the other type A (acute angled crystals, Fig.2)(Kihlström and Fjellström 1969 a, 1969 b). As seen from the figure the crystals are very similar to those occurring in dried samples of cervical mucus. As early as 1921 crystals were observed in semen from cattle, horse, and rabbit(Yamane 1921). Nevertheless, the occurrence of such crystals in semen has later been denied (Rydberg 1948, Rodriquez Villa 1958). It has recently been demonstrated that these crystals occur more regularly after removal of the sperm cells (Kihlström and Fjellström 1969 a). According to our observations crystals only occur exceptionally in whole semen and exclusively when there are very few spermatozoa present. Moreover, such crystals have been found in human seminal fluid (Kihlström and Fjellström 1969 a, 1969 b). In the rabbit at least the occurrence of these crystals is subject to a cyclic variation with a period length of about seven days (kihlström and Fjellström 1969 a, 1969 b). This length is close to that of the male sexual cycle in this species. Most likely a similar cyclic variation as to the pattern of crystallization will exist in bulls and men as well. Therefore, studies of these crystals in relation to the results of the inseminations will perhaps provide us with a tool for determining when a human male or a male animal possesses maximum fertilizing ability.

References

Bullough, W.S. Vertebrate Sexual Cycles. Methuen, London 1951

Degerman, G. and J.E. Kihlström, Acta physiol.scand. 1961. 51, 108

Degerman, G. and J.E. Kihlström, Acta physiol.scand. 1964. 62, 46

Degerman, G. and J.E. Kihlström, Acta physiol.scand. 1970. 78, 567

Doggett, V.C. Amer.J.Physiol. 1956. 187, 445

Doggett, V.C. and R.K. Keilers, Anat.Rec. 1962. 142, 227

Exley, D. and C.S. Corker, Biochem.J. 1965. 95, 54 P

Exley, D. and C.S. Corker, J.Endocrinol. 1966. 35, 83

Halberg, F., M. Engeli, C. Hamburger, and D. Hillman, Acta endocrinol. 1965. Suppl. 103, 1

Harris, G.W. Endocrinol. 1964. 75, 627

Hornstein, O. Personal communication 1965

Hornstein, O., J.E. Kihlström, and G. Degerman, Acta endocrinol. 1964. 46, 608

Inoué, S. Zool.Mag. Tokyo 1965 a. 73, 322

Inoué, S. Gunna Symp. Endocrinol. 1965 b. 2, 79

Järnebrand, K. and J.E. Kihlström, Acta physiol.scand. 1969. 75, 518

Kihlström, J.E. Studies in some activities of the male accessory glands, especially the production of male sperm antagglutin and their relations to fertility. Uppsala 1958

Kihlström, J.E. Arkiv Zool. 1962. 15, 359

Kihlström, J.E. Acta physiol.scand. 1963. 59, 370

Kihlström, J.E. Acta physiol.scand. 1965. 65, 61

Kihlström, J.E. and D. Fjellström, Acta obst.gynec.scand. 1969 a. 48, 147

Kihlström, J.E. and D. Fjellström, J.Reprod.Fert. 1969 b. 19, 375

Kihlström, J.E. and O. Hornstein, Acta endocrinol. 1964 a. 46, 597

Kihlström, J.E. and O. Hornstein, Z.vergl.Physiol. 1964 b. 49, 191

Kihlström, J.E. and C.A. Hultnäs, In manuscript 1970

Lisk, R.D. J.Exp.Zool. 1969. 171, 313

Månsson, J.C. Life Sci. 1965.4,329
Rodriquez Villa, L. Internat.J.Fert. 1958.4,42
Rydberg, E. Acta obst.gynec.scand. 1948.28,172
Tischner, M. Acta biol.Cracoviensia, Ser.Zool. 1967.10,283
Yamane, J. J.Coll.Agric.Hokkaido Imp.Univ. 1921.19,161

TRANSPLANTATION IN GYNECOLOGY

TRANSPLANTATION OF GYNECOLOGICAL ORGANS : ETHICAL PROBLEMS

L. Gedda

"G. Mendel" Institute for Medical Genetics

and Twin Research, Rome (Italy)

The transplantation of female reproductive organs poses biological as well as ethical problems. It is quite true that the former have already been considered by other speakers, but it must be stressed that they cannot be dissociated from the latter.

Science and technology keep introducing new ethical problems, especially in the medical field. We are not confronted with new ethics, but rather with new ethical problems. We are still largely in the experimental stage but, since man is involved, ethics ought to precede practice.

Individual problems ought to be viewed in the light of some basic principles:

<u>First</u>: primum non nocere, i.e. "first do not endanger" or "before all do no damage". This fundamental principle of ancient medicine is interpreted by Christianity and other religions as a consequence of the fact that life belongs to God and man is only entrusted with it. Thus the body is in-

tangible, unless any action be required for the superior health of the entire organism. Suicide and homicide are forbidden, as is any amputation not required by the "principle of totality".

Second: principle of totality (derived from natural law). This means that the health of the entire body has priority over the availability of individual organs. For instance, we may excise a cancerous organ, but not a healthy one. The same principle is to be applied to the availability of an organ for donation in case of transplantation. A twin may donate one kidney to his monozygotic cotwin because it has been shown that the remaining kidney may suffice for the needs of the body, while the principle of totality rules out the donation of organs that are not in duplicate.

It is quite clear that scientific knowledge is fundamental in ruling what falls inside or outside the principle of totality, and science may provide new perspectives for ethics.

Third: in other cases of transplantation, ethical problems concern donor and recipient. In gynecological transplantation we must also consider a third party: the potential progeny. This perspective concerns ethics and involves also public responsibility, since the state must protect public welfare as well as the health of the incapable.

Fourth: the right to exist has priority over the right to reproduce. In the case of a conflict between the desire of the woman to give birth to a child and the chance that the life of the child may be impaired, the right of the child prevails over the desire of the woman. This principle reflects a distinction between biological functions that may be renounced (such as reproduction) and biological functions that may not be renounced (because they are "quoad vitam").

Keeping in mind these principles, we should distinguish between generative organs sensu strictiori and generative organs sensu latiori (i.e., really reproductive organs and only operative organs respectively). In the former group I would place the ovaries, while the latter would include the tubes, uterus and vagina.

The allotransplantation of ovaries in the human, i.e. from one woman to another, should not be allowed because it may endanger the psychophysical health of the progeny, against the first, second, third and fourth principle. This type of graft effects a peculiar, biological type of adoption, yet it cannot alter the fact that the product of conception carries the genes of another woman. Thus in this case the recipient woman does not even "reproduce" herself according to the profound ethimological meaning of the word reproduction.

The transplantation of generative organs "sensu latiori", if biologically feasible, would seem to me to be ethically possible, if only the techniques involved could avoid endangering the product of conception, as immunosuppression could do especially if applied in loco. I believe we can rightly question the possible advantages of ovaries transplantation over legal adoption.

Yet here again we are faced with the problem of ethical objections against removal of organs that are not in duplicate, maiming the donor if living (against the principle of totality) often without guaranteed advantage for the recipient.

In any case much experimental proof would have to be gathered to establish beyond any doubt that:

- living donors would not suffer psychophysiological damage;

- control of recipient graft-rejection mechanisms would not interfere with the health of the product of conception;

- use of cadaver donors would not cause irreversible damage to gametes of recipient.

Several additional problems are common to other transplants, as indicated by Woodruff (1968), including the problem of an operative definition of death. But these general problems are secondary in the case of gynecological transplantation because here the first, fundamental, unsolved problem concerns the "third dimension": how to avoid damaging the product of conception.

The same Woodruff writes: "The doctor duty to his patient is not his only responsibility; there are many other people to be considered, including people who may be affected directly or indirectly by the patient's illness ..."; obviously the transplant is an artificial illness and fetus is an involved person.

Our task of M.D. is devoted, in any case and at any level, to helping and saving human life.

References

Gedda L. (1970). La legge naturale è veramente immutabile? In F. Angelini: Medicina e Morale. Vol. II, 27. Ed. Orizzonte Medico, Roma.

Gedda L., Milani-Comparetti M. (1969). An outline of the problems of immunogenetics and reproduction. Acta Genet. Med. Gemellol., 18:3-11.

Gruber A. (1969). Il principio di totalità e le sue applicazioni in medicina. In F. Angelini: Medicina e Morale, Vol. I, 96. Ed. Orizzonte Medico, Roma.

Lengerova A.L. (1969). Immunogenetics of tissue transplantation. North-Holland Publ., Amsterdam.

Woodruff M.F.A. (1968). The ethics of organ transplantation. In F.J. Ebling: Biology and Ethics. Academic Press, London.

MEDICAL ASPECTS ON TRANSPLANTATIONS IN GYNECOLOGY

Stig Kullander, M.D., Professor and Chairman of the Department of Gynecology at the University of Lund, Lund, Sweden

Ladies and Gentlemen

I was asked to join this panel to give briefly the view of a clinical man on transplantations in the gynecological field, preferably medical aspects.

I think we have the duty to fight for every single individual and therefore organ transplantation, seemingly a therapeutic alternative, is something in which we should interest ourselves. But it is a new field with several problems that should be dealt with without stirring up sensations or emotions.

There seems - as far as gynecological transplantations are concerned - to be no donor problems but the donation and the operation - a straight-forward technical problem with no moral trouble - must seem sensible and acceptable to the individuals. It must be individual decisisions by doctor, patient and donor (relatives as sister or mother, or other histocompatible women). In the selection of the recipient and the donor not the gynecologist alone but the team, with the immunologist and the anesthesiologist, should carry the responsibility. There should always be a great restraint in action.

I now would like to show some illustrating figures, 1 - 4.

FIG. 1. QUESTIONS CONCERNING TRANSPLANTATIONS IN GYNECOLOGY

1. Are uterus, tubes or ovaries rejected more indolently than other organs? (compare liver-kidney)

2. Is the new organ just surviving or functioning as well? (Responding to steroids; pregnancy?) (Ingrowth of nerves and lymphatics?)

3. What is the action of sex steroids on vascularization and immunological processes?

4. How about rejection crisis or infection? Could laparoscopy and hysteroscopy be used? Oedema and haemorrhages indications for removal?

5. Might the donor of uterus and/or Fallopian tubes be menopausal?

FIG. 2. OVARIAN TRANSPLANTATION

	Pieces to the rectus muscle		
	Embryonal (deep-frozen) ovarian tissue	Endocrine activity alone	
Ovaries			
	Diffusion chamber technique		
	Vascular anastomosis to ovarian vessels	To restore fertility as well	Indications: Aplasia, Oophorectomized women (tumor cases)
Suggestion	Experimental work on monkeys to study the reaction of the new ovary on gonadotrophins and during ALS- and immunosuppressive treatment		

FIG. 3. TRANSPLANTATION OF FALLOPIAN TUBES

Tube / and or without ovary	Vascular anastomosis to uterine and ovarian vessels	Indications: Tubes destroyed, occluded or taken away earlier
Suggestion	Experimental work on monkeys to study healing and function (peristalsis, ciliary function). Acute and chronic rejections.	

FIG. 4. TRANSPLANTATION OF UTERUS

Uterus (corpus) and or without tubes	Vascular anastomosis	Indications: Uterus destroyed, taken away or malformed (myomas, malignancies, malformations)
Suggestion	Experimental work on monkeys to study healing and function (endometrial response to steroids, growth during pregnancy, labor activity). Rejections.	

The Recipient should be less than 35 years and the physical and mental state of the patient should give promise of her being able to withstand the operation.

Probably, however, there is no direct surgical mortality. I make a reservation for hepatitis.

Immunosuppresive drugs (and ALS?) might be mutagenic. Therefore the patients should probably be given gestagens continously, in the beginning to prevent pregnancies and menstruations as well.

Antibiotics should be given to prevent infections.

Salpingectomy of a transplanted Fallopian tube as soon as the goal - pregnancy - is reached could prevent late rejections and complications and might be thought upon.

Regarding the Donor it makes it easier that ovaries and Fallopian tubes are double organs and not vital. Histocompatible donors, relatives who already have raised their families or healthy patients who are undergoing for example sterilization, should be used.

At last I would advocate that the first recipients should not be objects of experiments. An operation must really be necessary and must have some chances of success. First there then must be a sufficient number of promising animal experiments. Otherwise a transplantation can not be justified. Facts always must be explained to the patient and we should not give more hope to the patients than justified! But, of course, without trial there is in most cases no hope at all!

Before or parallel to human operations I would suggest much more experimental work, preferably on monkeys, being performed. It could, according to my opinion, be an effort directed by a small group of people interested in and working with transplantation problems in this field. This group could direct and coordinate research, they could arrange exchange of research-workers between different centers working on these problems and so on.

IMMUNODEPRESSION I. Principles

Gösta Magnusson

Surgical Department, Serafimerlasarettet, Stockholm

When an organ is transplanted into a recipient, who is genetically different from the donor, an immunological reaction - the so-called allograft reaction - is induced resulting in damage of the transplant (rejection). This process, shown in fig. 1 in a very simplified form, can be divided into many phases. The antigenicity of the allograft stimulates in a way not yet fully explained in an afferent arc the immune reacting centre (the regional lymph nodes, thymus and spleen) to production of proliferating immunized lymphocytes and to formation of antibodies against the antigens of the allograft. The efferent arc consists of transport of humoral antibodies and immunized lymphocytes to the allograft, where the allograft reaction is released. The cell-mediated immunity is probably most responsible to the allograft reaction.

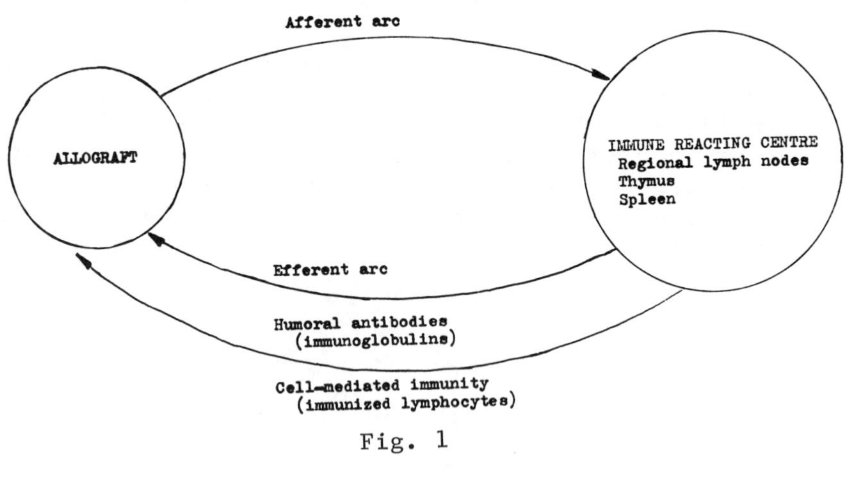

Fig. 1

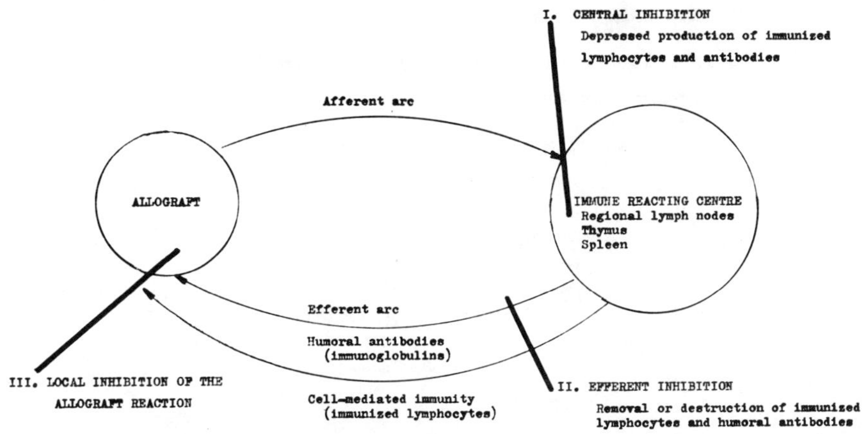

Fig. 2

The aim of the immunodepressive therapy is to stop the immunologic process at one or more of the different phases (fig. 2):
I. Central inhibition: Depressed production of immunized lymphocytes and antibodies,
II. Efferent inhibition: Removal or destruction of immunized lymphocytes and antibodies,
III. Local inhibition at the site of the allograft with the aim to reduce the effect of the allograft reaction.

Concerning the first principle, central inhibition, there are two practically available ways to develope this effect:

A. Reduction of the lymphoid tissue mass (fig. 3). This effect is recieved by surgical excision of lymphoid tissue, which produces immunized lymphocytes. Mainly thymectomy and splenectomy have been used. These methods seem, however, to be dubious, but in connection with other immunodepressive therapy they are perhaps of some clinical value.

I. CENTRAL INHIBITION

A. Reduction of the lymphoid tissue mass

Thymectomy

Splenectomy

Fig. 3

I. CENTRAL INHIBITION

B. Inhibited proliferation of immunized lymphocytes and formation of antibodies

1. Irradiation

General
Selective

2. Pharmacologic treatment

Steroids
Cytotoxic drugs
Antilymphocyte globulin

Fig. 4

B. The second way is to inhibit the formation of antibodies brought about by depressed proliferation of immunized lymphocytes (fig. 4). This effect can be developed by irradiation either generally of the total body or selectively of lymphoid tissue (regional lymph nodes, thymus and spleen). None of these methods are any longer used. - The most important part of immunodepression of central effect is laid on the pharmacologic therapy. Two types of drugs are mainly used: steroids and cytotoxic drugs. The mechanism of the steroid effect consists of depressed proliferation of immunized lymphocytes and in changing their immunologic potential. The cytotoxic drugs apparently act as antimetabolits by inhibition of the synthesis of nucleic acids. - Since 5 years antilymphocyte globulin has been used as a complement to the pharmacologic immunodepression. Its mode of action is still under investigation, but it is suggested that the proliferation of immunized lymphocytes is inhibited and that the circulating lymphocytes are destroyed.

The second principle in the immunodepression is removal or destruction of circulating immunized lymphocytes (fig. 5). Thoracic duct fistula produces effective removal of lymphocytes and humoral antibodies. By long-term lymph drainage (up to 40 days and more) the amount of circulating lymphocytes in the peripheral blood is

II. EFFERENT INHIBITION

A. Removal of immunized lymphocytes and humoral antibodies

Thoracic duct fistula

B. Destruction of circulating immunized lymphocytes

Irradiation of blood

Antilymphocyte globulin

Fig. 5

III. LOCAL INHIBITION AT THE SITE OF THE ALLOGRAFT

A. Irradiation of the allograft

B. Local steroid infusion

Fig. 6

depressed. The frequency of rejection chrises seems to be diminished by this therapy. - Prolonged irradiation of extracorporally shunted blood can produce lymphopenia. Usually the strong beta-emitting radioisotopes Co^{60} and Sr^{90} are used as irradiation sources.-Beside its central inhibitory mechanism, antilymhocyte globulin has a destructive effect on the circulating lymphocytes.

Finally, the third principle of the immunodepression is to protect the allograft from the immunologic attack of the immunized lymphocytes and humoral antibodies (fig. 6). Irradiation of the allograft has this effect, perhaps by destroying the immunized lymphocytes, which have infiltrated the allograft. - It is known that steroids have a stablising effect and therefore a protecting effect of the lysozome membrane of the allograft cells. This effect can be enlarged by local infusion of steroids into the artery of the allograft, because the potence of steroids is related to the administered dose.

About the effeciency of the immunodepressive methods here described it can be summerized: In cases of weak incompatibility the produced immunodepressive effect is enough, but in cases of strong incompatibility the immunodepressive effect is limited and only a little prolongation of the survival time is recieved. The methods here mentioned are unspecific and connected with side-effects, among which the depressed defence to infections is the most serious. The aim is to find more selectively acting immunodepressive methods for instance immunologic tolerance and immunologic enhancement.

IMMUNODEPRESSION II. Radiologic and pharmacologic treatment

Gösta Magnusson

Surgical Department, Serafimerlasarettet, Stockholm

The aim is to give a short review of practical immunodepression at organ transplantation by means of radiologic and pharmacologic treatment.

Central inhibition of the allograft reaction can be produced by irradiation either by general irradiation of the total body or by selective irradiation of lymphoid tissue (regional lymph nodes, thymus and spleen) (fig. 1). None of these methods are any longer utilized. - General irradiation in sublethal dose gives certainly an effective immunodepression but is connected with high frequency of irreversible damage of the bone marrow and bleedings difficult to treat. - Selective irradiation of lymph tissue has been found to produce an insufficient immunodepression. - Extracorporal irradiation of circulating blood administered in relative-

IMMUNODEPRESSION

A. Radiotherapy	
1. General irradiation	**Central inhibition**
2. Selective irradiation	**" "**
Regional lymph nodes	**" "**
Thymus	**" "**
Spleen	**" "**
Extracorporally shunted blood	**Efferent inhibition**
Allograft	**Local inhibition at the site of the allograft**

Fig. 1

IMMUNODEPRESSION

B. PHARMACOLOGIC THERAPY

1. Steroids (prednisone, prednisolone, 6-metyl-prednisolone)
2. Cytotoxic drugs
   Azathioprine
   Actinomycin C
3. Antilymphocyte globulin

Fig. 2

ly high doses gives inconstantly lymphopenia in peripheral blood. From the renal transplantation field some experience of this therapy has been recieved. The course of the irradiated patients is no different from that of patients given conventional immunodepressive drugs only, but the general impression is that the rejection episodes are less frequent and violent. - It is true that local irradiation of the allograft has been encouraging esspecially in renal transplantation. So a significant improvement of the survival rate by local irradiation of the allograft has been observed. As a routine in some transplantation centres the renal allograft is irradiated in connection with the transplantation, and this treatment is repeated in occasions of rejection.

The conventional immunodepression is mainly produced by pharmacologic therapy by means of steroids and cytotoxic drugs (fig. 2). - The basic steroid is prednisone. This treatment is started in connection with the transplantation with an individual dose of 50 - 300 mg prednisone. In the following course the aim is to reduce continuously the dose to levels high enough to prevent rejection and sufficient low to give no side-effects. The maintenance dose is 5 - 25 mg prednisone. The steroid treatment esspecially when given in high dosage (more than 25 mg prednisone) is hazardous.All side-effects of steroids described in the literature have been observed in our transplantation material. Full-moon face, adipositas, striae and skin echymosis are common symptoms. The depressed defence to infections is the most serious complication. Many of our patients have been lost due to fatal septicaemia. - Among the cytotoxic drugs the most important is a derivat of 6-mercaptopurine: azathioprine. Practical application of organ transplantation was possible first with the introduction of azathioprine. As distinguished from other cytotoxic drugs, azathioprine is an easily controllable substance. The azathioprine therapy is started before or in connection with the transplantation in doses of 2 - 3 mg per kilogram bodyweight. The maintenance dose, which is very

PHARMACOLOGIC THERAPY

Drug	Initial dose	Maintenance dose	Rejection dose
Prednisone	50 - 300 mg	5 - 25 mg	200 - 1,000 mg
Azathioprine	2-3 mg/kg	0,5-2 mg/kg	0,5-2 mg/kg
Actinomycin C			400 µg x III

Fig. 3

individual (from 1/2 mg to 2 mg per kilogram bodyweight), is held at a level high enough to give no side-effects in form of bone marrow depression, Therefore the picture of the peripheral blood must be checked regularely. Beside the blood dyscrasia, the toxicity of azathioprine appears in liver damage as cholestasis and in eruptions of mucous membranes. These complications can be so pronounced that temporary interruptions of the cytotoxic treatment must be done.

In occasions of rejection episodes the pharmacologic immunodepression is temporarely intensified. For a short interval steroids are given in high doses up to 1,000 mg prednisone and another cytotoxic drug, Actinomycin C, is added.

A survey of the dosage of the main drugs in different conditions is shown in fig. 3.

When there is a weak incompatibility between the donor and the recipient, pharmacologic therapy consisting of steroids and azathioprine in moderate doses produces an immunodepression sufficient to prevent rejection. In cases of strong incompatibility high doses of steroids together with other immunodepressive methods for instance removal of circulating lymphocytes, irradiation of blood and treatment with antilymphocyte globulin, can be necessary. Risks of serious side-effects of the immunodepression, especially fatal infections, increase. Even with this intensified immunodepression the allograft can be destroyed due to an irreversible rejection.

IMMUNOSUPPRESSIVE REGIMES DIRECTED AT THE LYMPHOCYTE

Carl G. Groth, M.D.

Department of Surgery, Karolinska Institute

Serafimer Hospital, Stockholm

According to present concepts, transplantation immunity is mediated by the small, immunocompetent lymphocytes; which cells are thymus dependent and recirculate via the lymphatic system. These concepts are supported by a body of experimental evidence which shows that immunological reactivity may be suppressed by thymectomy, drainage of lymph from the thoracic duct and the administration of specific antilymphocyte globulin. At some institutions these measures have been employed in association with clinical organ transplantation: The experience gained to date is outlined briefly in this paper.

Thymectomy. Removal of the thymus in connection with clinical organ transplantation was first performed in 1962 by Starzl and co-workers in Denver (12). For several of the patients, who received renal transplants from relatives, the postoperative course was remarkably smoth. These encouraging results prompted the Denver group, in 1964-66, to perform transthoracic thymectomy before transplantation in one half of their cases, while the other half served as controls. The donor-recipient lymphocyte antigen compatibility was similar in the test and control groups, and all the patients were given standard immunosuppressive drugs. At a follow-up study conducted 3 1/2 to 5 years later the thymectomy group was not found to display any advantage in terms of survival or quality of renal function. Histologic examination showed, however, that the homografts in the thymectomized patients had fewer and less severe lesions of a kind that would be expected to shorten the functional life of these organs (14).

In the last year thymectomy has also been performed in connection with heart transplantation (5), and at our Department the operation has been combined with drainage of the thoracic duct in recipients of renal grafts. It is, however, too early to assess the effect of this measure in these cases.

Thoracic duct fistula. Drainage of the thoracic duct lymph as an adjunctive immunosuppressive measure in clinical renal transplantation was first performed by Franksson (4) in Stockholm in 1964 and shortly after by Murray and co-workers (8) in Boston. In both cases the fistula was retained for a maximum of 2 weeks and in neither was any convincing improvement in the clinical course observed. Later on Fish and collaborators (2) and Franksson have cannulated the thoracic duct over longer periods and without simultaneous supply of the usual immunodepressive drugs. During treatment rejection was usually avoided but after closure of the fistula immunodepressive drug therapy had to be restored to.

The application of the method is restricted by the fact that in some patients cannulation is anatomically impossible, while in others the fistula ceases to function after a short time. Another disadvantage of the method is the large fluid and protein losses that must be replaced.

Anti-lymphocyte globulin. The serum of animals immunized with human lymphocytes contains antibodies to the lymphocytes in the globulin fraction. After purification, such antilymphocyte globulin (ALG) can be administered to patients by intramuscular or intravenous injection. The agent exerts an immunodepressive effect by destroying or inactivating the lymphocytes.

ALG was introduced in clinical renal transplantation in 1966 by Starzl (13). The addition of the agent to the standard immunodepressive drugs afforded a more effective prevention and treatment of rejection resulting in improved recipient survival and homograft function, despite the use of significantly smaller amounts of the classical drugs (11). These results have subsequently been confirmed by several groups in the United States and Europe (1,3,9, 15). Moreover, it has recently been established that ALG, given as the sole agent, exerts a definite immunodepressive effect in man, reflected in prolonged skin (9) and bone marrow (7) graft survival. In addition, ALG has proved to be an important prerequisite for successful liver transplantation, since these patients have a low tolerance of azathioprine (10).

Clinical ALG therapy has usually been limited to 2-4 months because longer periods increase the risk of undesirable reactions. These include fever, skin eruptions and anaphylactic reactions - that is, side effects that could be expected from heterologous protein therapy (6). The risk of Masugi or serum sickness nephritis is another limiting factor. Methods to prepare less toxic antilymphocyte globulins are, however, in progress.

Our knowledge of the mechanism of the immunological response to organ transplantation in man has been extended by the use of thymectomy, thoracic duct lymph drainage, and ALG. These measures have, in varying degrees, improved homograft function or survival. The results of ALG therapy can be expected to improve with the use of better preparations. Its combination, moreover, with one or both of the other measures may well prove valuable.

References

1. Brendel, W., Land, W.: Überraschende Ergebnisse durch intravenöse Therapie mit Antilymphozytenserum bei Organtransplantationen. Deutsch Med. Wschr. 93:2309, 1968.

2. Fish, J.C., Sarles, H.E., Tyson, K.R.T., Remmers, Jr., A.R., Ritzmann, S.E.: The immunologic consequences of lymph lymphocyte depletion. Surg. Forum 20:268, 1969.

3. Franco, D., Hardy, M.A., Quint, J., Agarwal, B., Lathum, W., Monaco, S.T. and A.P.: Use of a DEAE-cellulose-human erythrocyte stroma column to produce anti-human lymphocyte globulin devoid of thrombocytopenic toxicity: method and preliminary results. Europ. Surg. Res. 1:213, 1969.

4. Franksson, C., Blomstrand, R.: Drainage of the thoracic lymph duct during homologous kidney transplantation in man. Scand. J. Urol. Nephrol. 1:123, 1967.

5. Kahn, R., Reynolds, Jr., E.W., Walton, Jr., J.A., Kirsh, M.M., Vathayanon, S., Sloan, H.: Human heart transplantation for cardiomyopathy. Surgery 67:122, 1970.

6. Kashiwagi, N., Brantigan, C.O., Brettschneider, L., Groth, C.G., Starzl, T.E.: Clinical reactions and serologic changes after the administration of heterologous antilymphocyte globulin to human recipients of renal homografts. Ann. Intern. Med. 68:275, 1968.

7. Mathé, G.: Approaches to the immunological treatment of cancer in man. Brit. Med. J. 4:7, 1969.

8. Murray, J.E., Wilson, R.E., Tilney, N.L., Merrill, J.P., Cooper, W.C., Birtch, A.G., Carpenter, C.B., Hager, E.B., Dammin, G.J., Harrison, H.J.: Five years' experience in renal transplantation with immunosuppressive drugs: survival, function, complications, and the role of lymphocyte depletion by thoracic duct fistula. Ann. Surg. 168:416, 1968.

9. Najarian, J.S., Simmons, R.L., Gewurz, H., Moberg, A., Merkel, F., Moore, G.E.: Anti-serum to cultured human lymphoblasts: preparation, purification and immunosuppressive properties in man. Ann. Surg. 170: 617, 1969.

10. Starzl, T.E. with the assist. of Putnam, C.W.: Experience in Hepatic Transplantation. W.B. Saunders Company, Philadelphia, 1969.

11. Starzl, T.E., Groth, C.G., Terasaki, P.I., Putnam, C.W., Brettschneider, L., Marchioro, T.L.: Heterologous antilymphocyte globulin, histoincompatibility matching, and human renal homotransplantation. Surg. Gynec. Obstet. 126:1, 1968.

12. Starzl, T.E., Marchioro, T.L., Talmage, D.W., Waddell, W.R.: Splenectomy and thymectomy in human renal homotransplantation. Proc. Soc. Exp. Biol. Med. 113:929, 1963.

13. Starzl, T.E., Marchioro, T.L., Porter, K.A., Iwasaki, Y., Cerilli, G.J.: The use of heterologous antilymphoid agents in canine renal and liver homotransplantation and in human renal homotransplantation. Surg. Gynec. Obstet. 124:301, 1967.

14. Starzl, T.E., Porter, K.A., Andres, G., Groth, C.G., Putnam, C.W., Penn, I., Halgrimson, C.G., Starkie, S.J., Brettschneider, L.: Thymectomy and renal homotransplantation. Clin. Exp. Immun. 6:803, 1970.

15. Traeger, J., Carraz, M., Fries, D., Perrin, J., Saubier, E., Bernhardt, J.P., Revillard, J.P., Bonnet, P., Archimbaud, J.P., Brochier, J.: Studies of antilymphocyte globulins made from thoracic duct lymphocytes. Transpl. Proc. 1:455, 1969.

TRANSPLANTATION OF OVARIES

Göran Möller

Department of Bacteriology, Karolinska Institutet

Stockholm 60, Sweden

Tissue grafts exchanged between different individuals of the same species are rejected as a consequence of an immune reaction. If the individuals are genetically identical there is no rejection. This situation exists only between monocygotic twins or between members of an inbred animal strain. However, graft rejection may also fail due to particular characteristics of the grafted tissue. Thus, cartelage will survive in histoincompatible recipients due to its unusual resistance to homograft rejection. Even highly susceptible tissues may survive if they are grafted into priviliged sites, the most well-known of these sites being the entire chamber of the eye.

The immune reaction leading to graft rejection can be specifically abolished by two different phenomena: 1. immunological tolerance and 2. immunological enhancement. Tolerance represents a specific inactivation of the immunological system to one particular antigen after administration of the antigens under certain circumstances. Thus, tolerance can be induced by treating neonatal animals with histoincompatible cells. When the animals grow up they have lost their ability to reject graft from the cell donor, but have an intact ability to reject grafts from all other individuals. Tolerance is not restricted to the newborn period, but can also be achieved in adults, although larger antigen doses are usually required.

Immunological enhancement is an entirely different phenomenon. Enhancement also represents a specific failure of the homograft reaction, but is mediated by humoral antibodies directed against the graft. Thus, donors pretreated with antibodies directed against a foreign graft may accept grafts from this individual, but reject other grafts in a normal way. Enhancement is more restricted than tolerance in its applicability. Thus, tumors are particularly susceptible to enhancement wheres most normal tissues are difficult to enhance by humoral antibodies.

Ovaries have been transplanted for a long time period and successes and failures as well, have been reported. However, most of these old findings are not particularly illuminating, because the experiments were performed in non-inbred animals. When inbred animals become available and when the knowledge of the immunological reactions causing graft rejection had increased, a removed interest in ovarial transplantation occurred. It was found that ovaries, like all other normal and neoplastic tissues, were regularly rejected, if they were grafted between individuals being different from each other with regard to the major transplantation system of the species. It has been found that there is one dominant system of transplantation antigens, which causes a very marked graft rejection after 8–11 days if grafts are exchanged between individuals being different with regard to this system. However, in addition to this dominating system there is a number, probably 30 or more, of so called weaker systems. Grafts exchanged between individuals being different only with regard to the weak systems are also rejected, but after a considerably longer time. When ovarial transplants were performed between individuals which differred only with regard to weak antigens it was found that the grafts survived indefinitely in a number of situations,in contrast to all other normal tissue studied (Linder, 1961). This was an unexpected result and clearly distinguished ovaries from skin and other normal tissues. The immunological mechanism behind this increased survival was studied in a later report (Linder, 1962 a,b). It was found that ovaries not only were resistant to graft rejection across minor histocompatibility barriers, but also changed the reactivity of the host to other normal tissues of the same strain. Thus, animals which had been grafted with ovaries, were subsequently found to be unable to reject skin homograft, which are invariably rejected into non-ovarial grafted hosts. The kinetics of this change of the host was complex. Thus, if skin grafts were performed between 10 to 70 days after the ovarial graft, the skin grafts were rejected in an accellerated way. Obviously the ovarial transplant had sensitized the host to such an extent that a subsequent skin graft was destroyed more rapidly. In spite of this the survival of the ovary itself was not impaired indicating that the overy is highly resistant to the homograft reactivity. However, if the ovarial graft has been in place for more than 70 days, a subsequent skin graft was not rejected at all, but showed indefinite survival. In order to induce this unresponsiveness ovarial grafts had to be persistent all the time in the host. Thus, if the graft was removed after various intervals (90–210 days) in the host, the treatment failed to prolong the survival time of subsequent skin grafts significantly. It was furthermore demonstrated by Linder (1962a) that the unresponsiveness was specific, i.e. it was only extended to skin grafts derived from the same strain as the ovary; other skin grafts being rejected in the normal time.

There are only two possible interpretations of these results, 1. The ovarial graft induced a state of immunological tolerance during its residence in the foreign host, 2. The graft stimulated production of enhancing humoral antibodies, which were responsible for the increased survival time of the subsequent skin graft. Attempts were made (Linder, 1962b) to distinguish between these possibilities. He found that transfer of lymphoid cells from untreated animals into animals Iretreated with an ovarial graft abolished the induction of unresponsiveness to the skin grafts. This would suggest that the animals grafted with ovaries had a defect in their lymphoid cell population, making them unable to reject the subsequent skin graft, but that this defect could be repaired by the adoptive transfer of normal lymphoid cells. These findings would be in agreement with the interpretation that unresponsiveness is due to immunological tolerance, because this phenomenon is known to involve a central inhibition of the immune response. Furthermore, attempts to detect humoral antibodies directed against the donor´s tissue, as would be required for calling the phenomenon immunological enhancement, failed entirely.

Nevertheless, immunological enhancement can be easily demonstrated with ovarial grafts in contrast to most other tissues. Thus, it was found (Möller, 1964) that animals pretreated with humoral ant ibodies directed against the genotype of the ovarial graft donor caused a marked prolongation of the survival times of the grafted ovaries, some of them surviving indefinitely. Analogous experiments were performed with skin grafts. In these experiments only a limited prolongation of graft survival time maximally ranging to 3 days, were observed. Thus, ovaries, like tumors were found to be unusually susceptible to the induction of the immunologically enhancement and could survive in antiserum-treated mice for very long time periods. Taken together the finding indicate that ovaries are unique from an immunological point of view. Thus, they will survive by themselves in genetically dissimilar animals, provided that the discrepancy between donor and host does not involve the major histocompatibility locus. They are also competent to change the host in such a way that subsequent skin grafts are permanently accepted. Although the actual mechanism of this ability of ovarie to change the host reactivity is not known, most of the data suggest that it is due to induction of immunological tolerance. The mechanism for tolerance induction is presently unknown, but it seems plausible to assume that the large amount of cystic fluid produced by the transplanted ovaries represents a source of soluble transplantation antigens responsible for induction of tolerance. It has been demonstrated that the fluid collected from such cysts contains large amounts of histocompatibility antigens in a soluble form and it is well known that soluble antigens favours induction of tolerance. Furthermore, ovaries were unusually susceptible to induction of immunological enhancement by humoral antibodies, in contrast to other normal tissues, such as skin grafts.

From animal experiments performed so far it would appear that the ovary represents an uniquely suitable material for transplantation between genetically dissimilar individuals,provided host and donors do not differ at the major histocompatibility locus. In view of the present knowledge of histocompatibility antigens in human, it seems likely that host and donor can be matched with regard to the major HL-A systems so that only a transplantation is performed between identical individuals. This situation is most easily met when grafts are performed between siblings where 25% are identical at the major locus. It can be expected that in such a case the grafts may function in a large proportion of cases for a very long time period, even in the absence of immunosuppressive treatment, as is necessary with other types of tissue grafts.

SUMMARY

It has been demonstrated that ovaries can survive in incompatible recipients provided that the incompatibility does not concern the major locus within the species. The ovaries also change the reactivity of the host, so that other normal tissues are accepted, presumably because the ovarial graft has induced immunological tolerance. Ovaries are also susceptible to the induction of immunological enhancement by pretreating the host with humoral antibodies directed against the ovary. Taken together, the findings suggest that allogeneic ovaries grafts may be more likely to succeed than other normal tissues.

REFERENCES

1. Linder, O. (1961). J. Nat. Cancer Inst. 27, 351.
2. Linder, O. (1962a). Immunology 5, 195.
3. Linder, O. (1962b). Thesis, Balders tryckeri, Stockholm, Sweden.
4. Möller, G. (1964). Transplantation 2, 281.

SIGNIFICANCE OF TRANSPLANTATION OF THE UTERUS AND THE FALLOPIAN TUBES

P. Vara, M. Seppälä and S. Paatsama

Department II of Gynaecology and Obstetrics

University Central Hospital, Helsinki, Finland

INDICATIONS

So far, uterine and adnexal transplantations have their significance in the experimental research, where the utero-ovarian cyclic function and the immunological phenomena of a fertilized egg in its new surroundings are of interest. The uterus is capable of harbouring the foetus, which is genetically only half-maternal of origin. If a successful transplantation of the uterus and the fallopian tubes is followed by pregnancy, the transplanted uterus is totally foreign to both the host and the zygote. It is difficult to predict the outcome of such a pregnancy. Even from a very optimistic point of view, hopeless cases of sterility could be the only indication of uterine and tubal transplantations in the future human medicine.

PROBLEMS OF UTERINE AND TUBAL TRANSPLANTATIONS

As in any organ transplantation, ethical and juridical problems arise in providing donors of healthy organs. If ovaries are also transplanted, the ovum and the following zygote will be totally foreign to the recipient, who actually cannot have a child of her own genetic make-up.

Immunological problems are still the major obstruction in all tissue homotransplantations. An additional problem arises, if pregnancy occurs in a transplanted uterus. The immunosuppressive anti-mitotic treatment of the host's immunological reaction must not be harmful to either the foetus or to the remaining ova, which might later be fertilized.

A close histocompatibility between the donor and the recipient is necessary to minimize the host's immunological reaction. Therefore tissue typing is necessary. Management in tissue preservation is also important, but today it is still very poorly known.

Technical problems may be more easy to overcome. Surgical management requires refined vascular techniques. Prevention of microthromboses and knowledge of the minimum nutritional requirements of the genitals are further problems.

OWN EXPERIMENTS

The nutritional requirements of the uterus, the fallopian tubes and the ovaries were studied in a 7 year old Pithecus Rhesus. Exstirpation of the uterus with its adnexa was performed in such a way that only the vascular connection of the right uterine artery and vein was left open. The uterus was then placed back to its anatomical location by chromic catgut sutures without any other vascular connections. The recovery of the ape was normal, and she had one vaginal bleeding 14-16 days after the operation. After 40 days had passed from the operation the animal became aggressive. There was no further vaginal bleeding. The autopsy was performed 60 days after the operation. The macroscopical outlook of the uterus, the fallopian tubes and the ovaries was normal. In the microscopic examination (Prof. L. Hjelt) the endometrium was found to be at a proliferative stage. The myometrium and the ovaries had a normal structure. Slight degeneration of connective tissue was observed in the fimbrial part of the right fallopian tube.

The effect of immunosuppressive treatment may not always be harmful to the growth of the foetus. This is indicated by a case of a 22 year old woman, who had received a kidney from her brother. She became pregnant two years after the transplantation, but concealed her pregnancy for five months in a fear of medical abortion. During this time she had periods of immunosuppressive treatment (azathioprine 5 mg/kg/day and prednison). The child looked quite normal out, weighed 3330 g and had 10 Apgar points. The gestational age at the time of delivery was 38 weeks, and the weight of the placenta was 470 g. The macroscopic and histologic structure of the placenta was normal.

DISCUSSION AND CONCLUSIONS

The nutrition of the uterus may even be maintained by the omentum alone. Bykov (1927) resected the uterus free of its blood supply and wrapped it in the omentum in dogs. The uterus was found to be viable after eight months on exploration. By a vascular anastomosis the uterus with its adnexal organ was autotransplanted in 18 dogs (Eraslan et al. 1966). 3 of the dogs became pregnant and 2 delivered. A successful homotransplantation of the uterus and the ovary was performed by vascular anastomosis in 14 dogs (Yonemoto et al. 1969). 5 of them had viable uteri 45 days later at the autopsy. According to our experiment it seems that enough nutrition for the genitals can be supplied by one uterine artery and one vein. The nutrition might be improved, if the uterus were wrapped in the omentum. The vaginal bleeding, which occurred 2 weeks after the operation, may represent a menstruation bleeding, but it may as well have been some kind of a withdrawal bleeding. Although the organs appeared normal 60 days after the operation at the autopsy, it is difficult to evaluate the final functional capacity of such autotransplanted genitals, because these animals are poorly fertile outside their natural surroundings.

Such a developing and growing element as the foetus is, it could be affected by the immunosuppressive antimitotic treatment, especially at the stage of organogenesis. However, in our case no harmful effect could directly be seen. The development of the child and his immunological and reproductive systems remains to be seen in the future.

Today we have a strong feeling that transplantation of the uterus and the fallopian tubes may never have any large scale significance in the human medicine, but it may get some importance in the future in the field, which might be called "high society medicine".

ANASTOMOSIS OF SMALL VESSELS

Argo Kövamees
Dept. of Surgery
Serafimer Hospital
Karolinska Institute, Stockholm

Poiseuilles law states that - under controlled conditions - the flow through a vessel is a function of the fourth power of its radius. "In a geometric progression of this type minute changes in diameter have a profound effect on flow" (Jacobson, 1967).

The constriction caused by the suture line following surgery of large and medium sized arteries does not usually critically alter blood flow. An end-to-end anastomosis is usually performed with two continuous over-and-over sutures of atraumatic 5-0 or 6-0 synthetic monofilament material on each side. In smaller vessels, however, as the case may be in transplantation of gynecological organs, flow problems become more significant. According to Eastcott the critical diameter is 5 to 6 mm-s: arteries of smaller diameter can often be reconstructed by conventional techniques but the results are not always good. We have encountered this problem in reconstruction of arteries in the lower leg. In cases of marked narrowing of the lumen a "patch graft" of vein is inserted over the incision.

For small vessel anastomosis new suture materials and instruments have recently been developed (Jacobson, 1967). Ultrafine suture and needle with a diameter of 0.005 inches with a 0.001 inch monofilament nylon suture may be used. The conventional suture technics with somewhat finer instruments can, however, be used without help of any special apparatus. Real problems come when small vessels as those of rabbit´s ear, have to be anastomosed (Bellman, 1953). For that microsurgical instrumentation with operation microscope, pneumatic instruments etc. are needed, and this field is still rather experimental and the work too time consuming yet for use in transplantation surgery at time being.

To achieve technical perfection and to cut the time needed for small vessel anastomosis, various suturing machines have been constructed. Androsov's machine (fig. 1) has more than thirty parts.

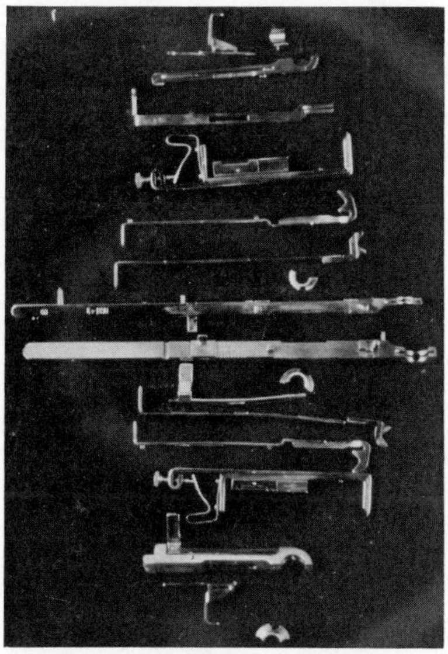

Fig. 1. Main parts of Androsov's stapling machine.

It staples the evaginated ends together (fig. 2 A) and turns one end over the anastomosis line (B). When a vessel is arteriosclerotic the intima is rather stiff and it may fracture when everted and a flap may broke loose from the intima at the line of anastomosis (Bellman et al., 1967).

A non suture small vessels prosthetic connector has been developed by Selker and Carney, New Haven, 1967. The connector is fabricated of stainless steel in the form of a ring with an internal diameter measuring either 1.5 or 2.0 mm. It is put in place with the help of a stabilizing arm, which can be grasped by a simple hemostat. The connector is slipped over a guide wire so that all is in readiness when the vessel is clamped and the connector is placed over the afferent end of the vessel. Using the guide suture the vessel is evaginated over the top of the ring and the efferent end is brought over the top of the so evaginated afferent end and likewise impaled on four prongs.

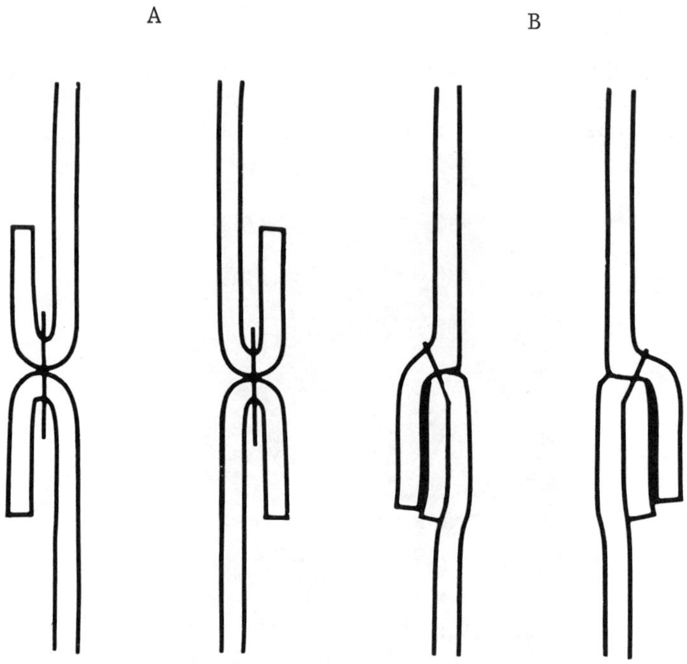

Fig. 2. The principle of anastomosis with Androsov's stapling machine. The ends of the vessels are everted and united with minute clips (A) . In a second stage the vessel ends are turned over one another (B).

With Nakayama's apparatus (fig. 3) the vessels are anastomosed with two metal rings put together with two ring holding clamps (fig. 4). Each ring has six evenly spaced pinholes and interposed between the holes are six pins. A ring is slipped over each end of the vessels to be joined. The edge of the vessel is then everted on the surface of the ring and hooked by the pins. The two rings are brought together so that the pins of each ring enter the holes of the opposing ring. We have found Nakayama's instrument more easy to handle than the Androsov's apparatus. It is considerably less traumatic, since the vessels do not have to be completely everted. Fig. 5 shows an autopsy specimen with Nakayama's ring and fig. 6 shows a transplanted segment cut longitudinally. Anastomosis with Nakayama's instrument can be done both end-to-end and end-to-side (fig. 7).

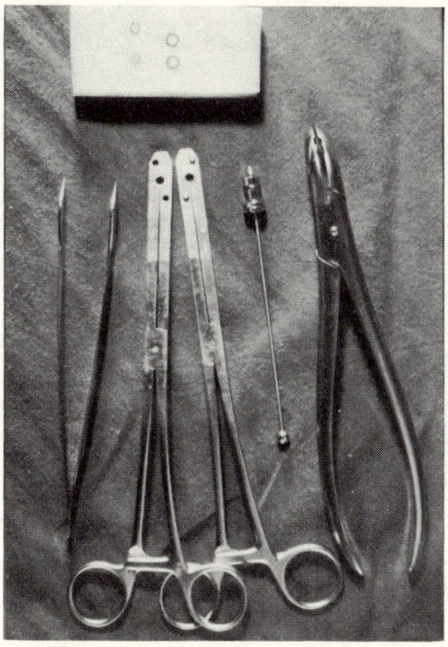

Fig. 3. Nakayama's instrumentation.

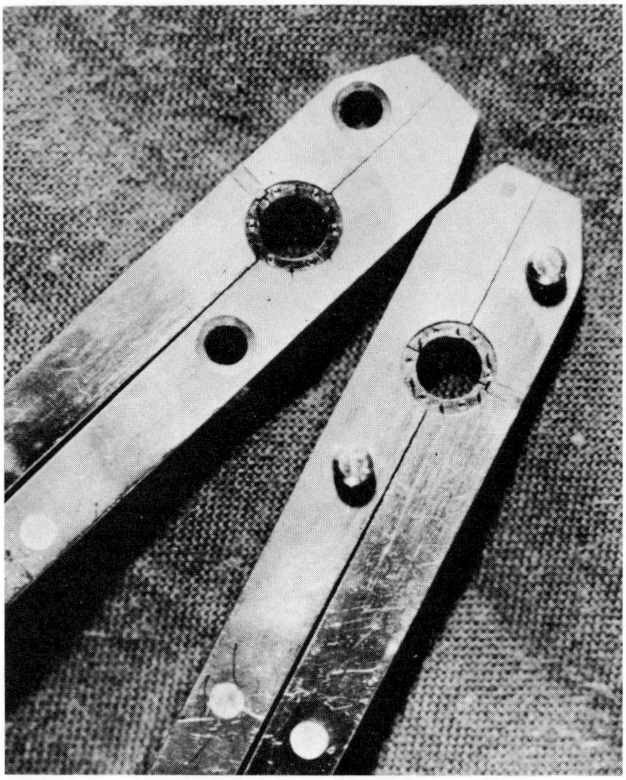

Fig. 4. Nakayama's ring holding clamps with rings inserted, ready to be slipped over each end of the vessels to be joined.

Fig. 5. An autopsy specimen with Nakayama's ring.

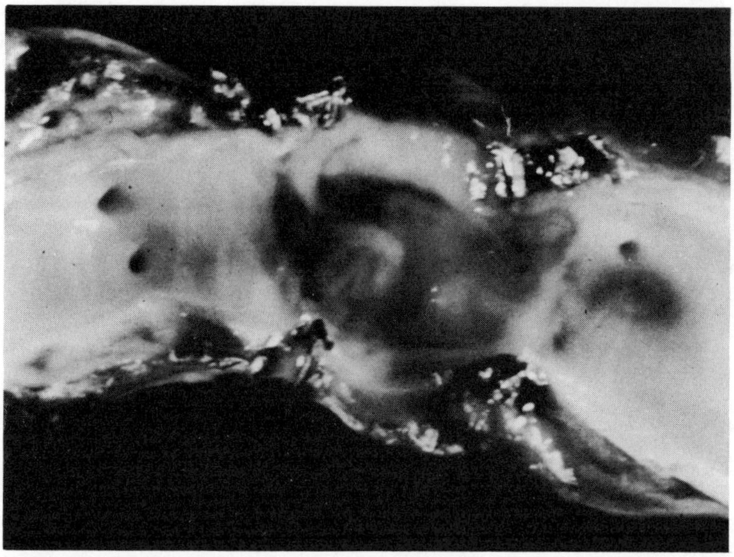

Fig. 6. An autopsy specimen with the transplanted segment cut longitudinally and two rings transsected.

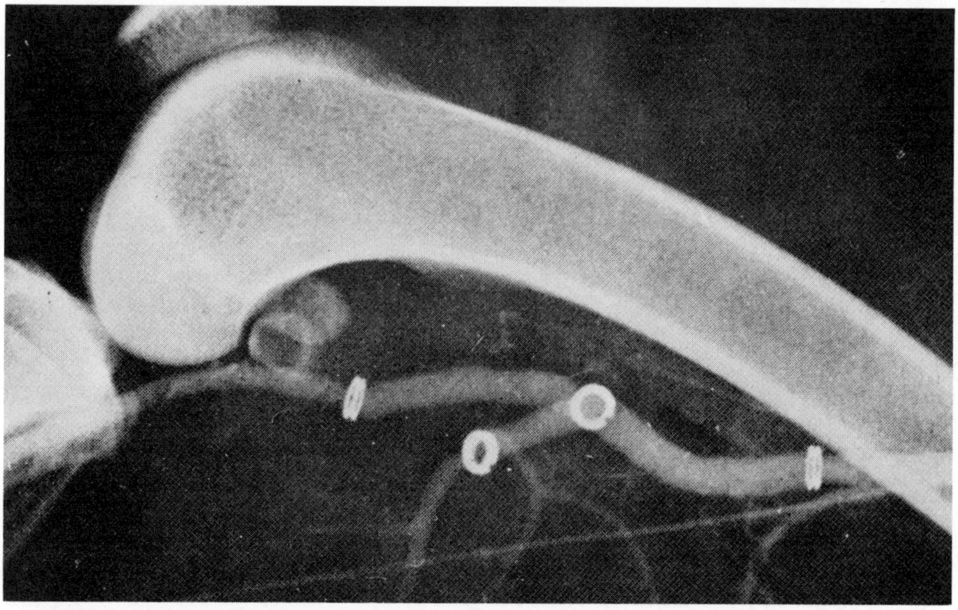

Fig. 7. Angiography of an anastomosis in dog with an end-to-end transplanted vein graft and an end-to-side transplanted graft anastomosed to it. (Bellman et al., 1967).

Concluding: We have found that anastomosis of small vessels can give good results even with conventional surgical techniques, when constriction of the lumen can be avoided. Of stapling machines we found Nakayama's instrument easier to handle than the Androsov's instrument, but all stapling machines in general need a lot of practice to handle. They make, however, vein grafting easier when the diameter of the lumina differs considerably. A special problem is measuring of blood flow after anastomosis of small vessels. We have used the radioactive ^{133}xenon gas tissue clearence technique by local injection and external counting.

BIBLIOGRAPHY

Androsov, P.I.: Innovations in vascular surgery. New Surgical Equipment and Instruments and Experiments in their Use, p. 75. National Science Foundation, Washington, D.C., 1961 (1957).

Bellman, S.: Microangiography. Acta Radiol. 1953, Suppl. 102.

Bellman, S., Kövamees, A. & Rietz, K.-A.: Reconstruction of Small Arteries: Androsov's and Nakayama's Apparatuses and Microangiography. In Donaghy, R.M.P. & Yaşargil, M.G.: Micro-Vascular Surgery, p. 67-75, Thieme, Stuttgart, 1967.

Jacobson, J.H.: The Development of Microsurgical Technique, p. 4-14 in Micro-Vascular Surgery, Thieme, 1967.

Kövamees, A.: Skin Blood Flow in Obliterative Arterial Disease of the Leg. Effect of Vascular Reconstruction Examined with Xenon and Iodine Antipyrine Clearance and Skin Temperature Measurements. Acta chirurg. scandinav. suppl. 397, 1968.

Nakayama, K., Tamiya, T., Yamamoto, K. & Akimoto, S.: A simple new apparatus for small vessel anastomosis (free autograft of the sigmoid included). Surgery, 1962, 52:918.

Selker, R.G. & Carney, P.M.: A Non-Suture Small Vessels Prosthetic Connector. p. 50-52, Micro-Vascular Surgery, Thieme, 1967.

Yasargil, M.G.: Microsurgery - Applied to Neurosurgery. Thieme, Stuttgart, 1969.

AUTOTRANSPLANTATION OF THE UTERINE HORNS AND OVARIES IN FEMALE RABBITS

Macků F., Kučera E., Tachezy R., Studený B., Trnka V.

2nd Dep. of Obst.Gyn.,Charles University, Prague

At our Clinic we were already previously concerned with the problem of the possible transplantation of female sex organs, as apparent from the lectures and papers of one of our collaborators on the experimental transplantation of the pituitary, tubera cinerea and ovaries. In recent years we rather reflected on the transplantation of the ovaries and oviducts. These operations could be of therapeutic use in irreparable obstructions of the oviducts or conditions after bilateral salpingectomy.

In order to test this procedure experimentally we began in 1968 with autotransplantations of the uterine horns and uterine horns and ovaries in female Chinchilla rabbits with an average weight 3 500 g. After opening the abdominal cavity we ligatured the vessels supplying the uterine horns and on each side we removed a 5 - 6 cm long portion of the horn, rinsed it in a solution of Pencillin and Streptomycin and implanted them on the contralateral side / Figs. 1,2 /. For the suture we used fine catgut in an atraumatic needle above a polyethylene catheter which to this end was introduced by incision in the joint portion of both horns. When the suture was completed, the transplant was sutured to the serous membrane of the plica lata, the catheter was removed and the incision made for it closed by catgut stitches. / Figs. 3,4/.

Before closing of the abdominal cavity we instilled by the intraperitoneal route Kanamycin which was administered also for three day after operation by the intramuscular route. In several

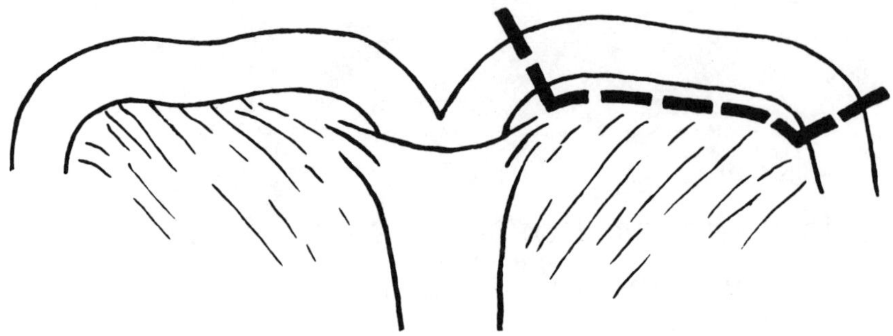

Fig. 1

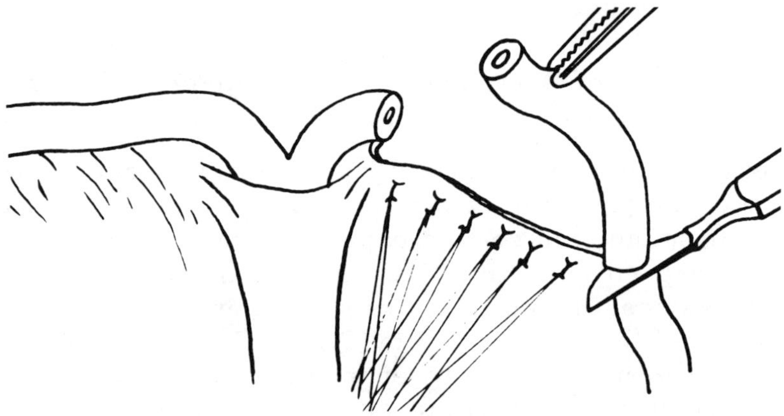

Fig. 2

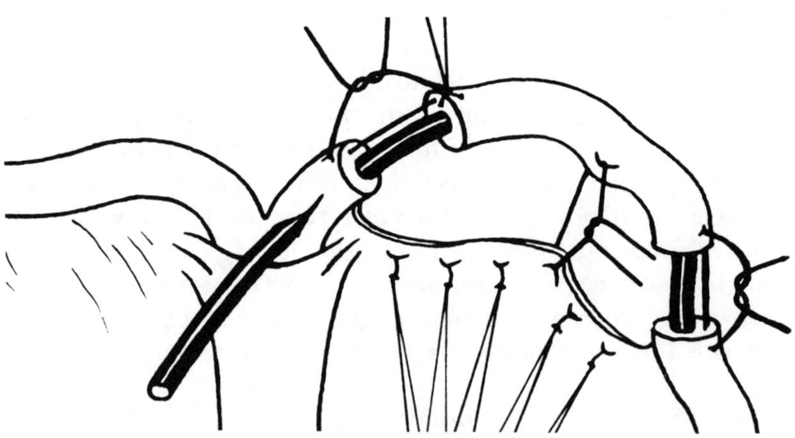

Fig. 3

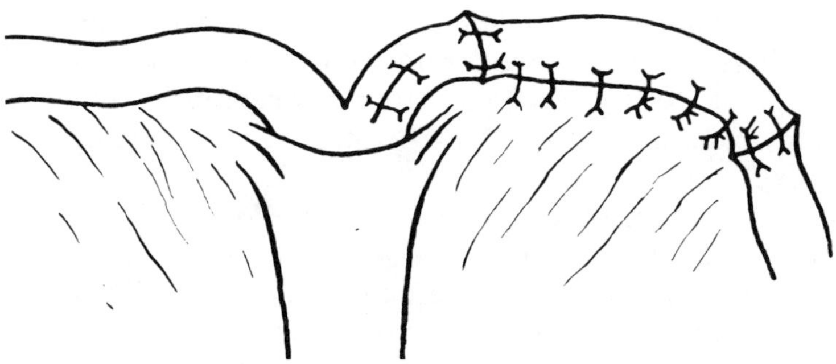

Fig. 4

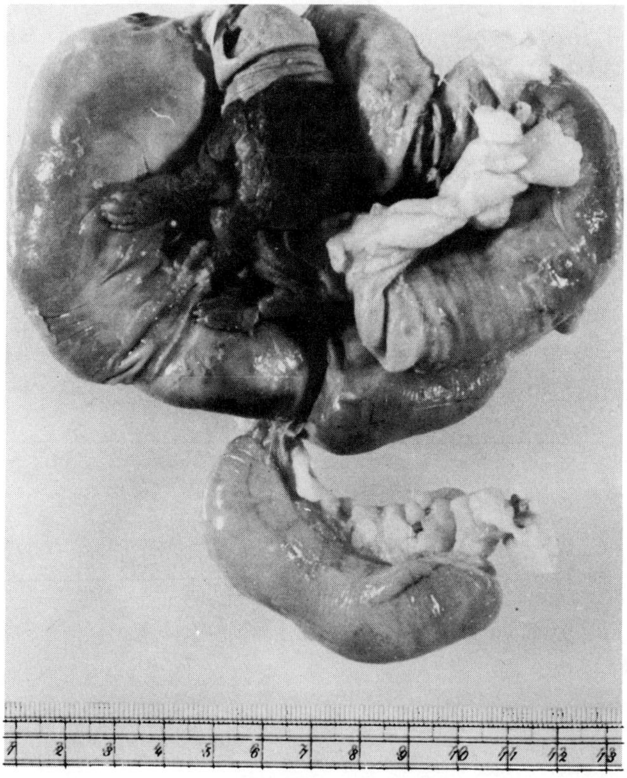

Fig. 5

instances we transplanted thus the entire peripheral section of the uterine horn incl. the oviduct and ovary.

We investigated the incorporation of the transplant in the early stage after operation. In the present paper we are reporting on the results of the macroscopic finding during a check-up / relaparotomy/ made 6 - 8 weeks after operation. The results of microscopic examinations and functional investigations will be submitted later.

A total of 30 transplantations was performed in 15 rabbits. In about one third, probably as a result of occlusion of the central suture, an expansion of the centrally located transplant with a serous or seromucinous content occurred. Two animals died within 24 hours after operation from peritonitis. Twothirds of the transplants were incorporated. Fig. 5 illustrates the results of transplantation when two months after operation the female was pregnant.

We feel that these results justify the continuation of our experiments. Although the operation is not particular pretentious from the technical aspect, the procedure must be improved further and more favourable conditions must be created for the vascularization of transplants. We began therefore with experimental homotransplantations.

HOMOTRANSPLANTATION OF THE UTERINE HORNS IN FEMALE RABBITS

Macků F., Kučera E., Tachezy R., Studený B., Trnka V.

2nd Dep. of Obst. Gyn., Charles University, Prague

When we found that it is possible to transplant part the the uterine horns from one side to the other in the same animal, we began to transplant the uterine horn from one animal to another.

Two surgical teams worked concurrently. The same procedure was used as described in the evaluation of autotransplantations: after ligature of the supplying blood vessels we removed segments of the uterine horns 5 - 6 cm long, their central portion was marked by a catgut stitch. After rinsing in a solution of antibiotics the transplant was handed to the other team for processing. A central and peripheral suture end-to-end was performed above an inserted polyethylene catheter. Then the transplant was stitched to the plica lata, the catheter was removed and the incision made for the catheter was closed. Before suture of the abdominal wall Kanamycin was instilled by the intraperitoneal route and was administered for another three days after operation by the intramuscular route.

In this way we exchanged mutually parts of bo th horns of the uterus in six pairs of female rabbits /Chinchilla, average weight 4000 g /. One animal died during operation, two the second day after operation from peritonitis. Two to three weeks after operation relaparotomy was performed in the remaining nine animals and the results of 18 transplantations were subjected to macroscopic examination:

- 4 times the transplant did not take

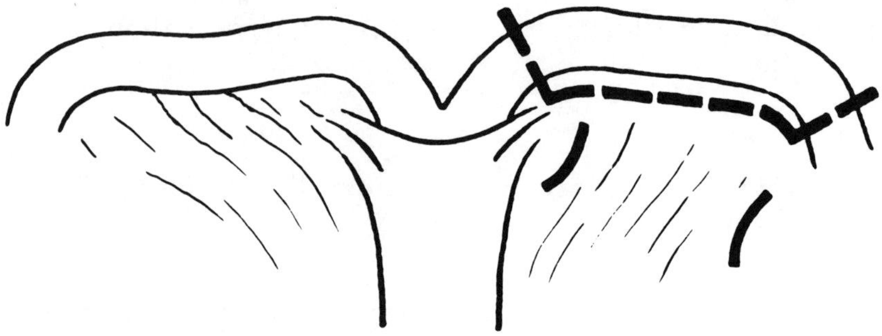

Fig. 1

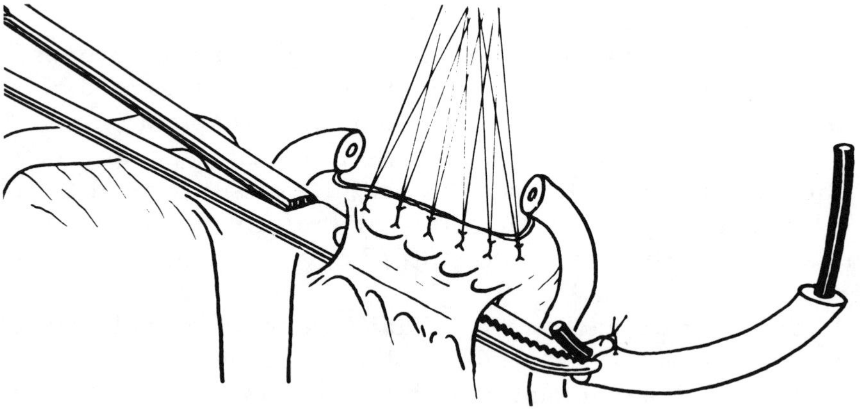

Fig. 2

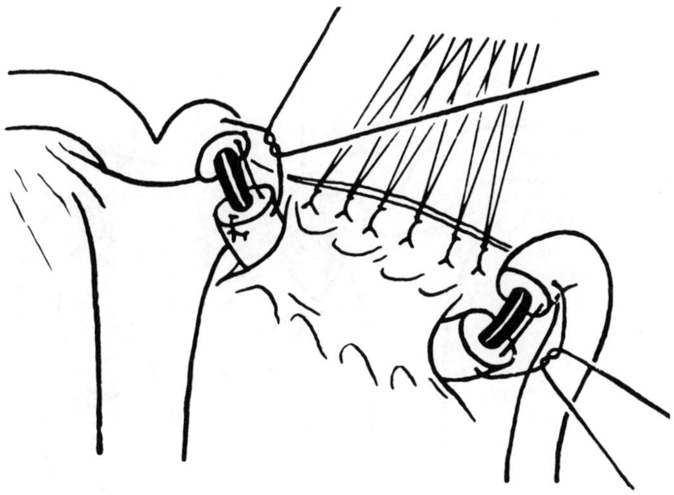

Fig. 3a

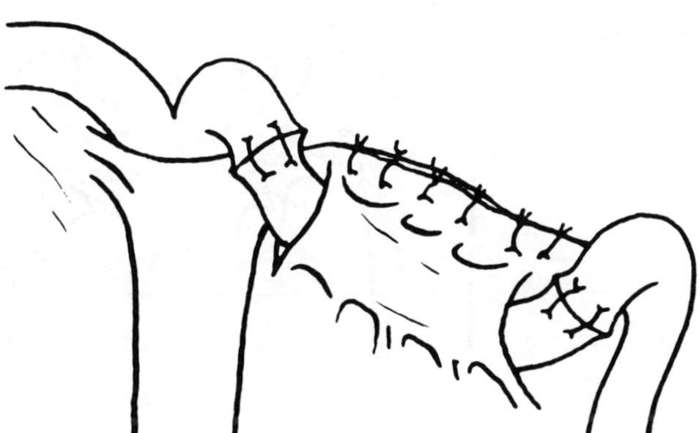

Fig. 3b

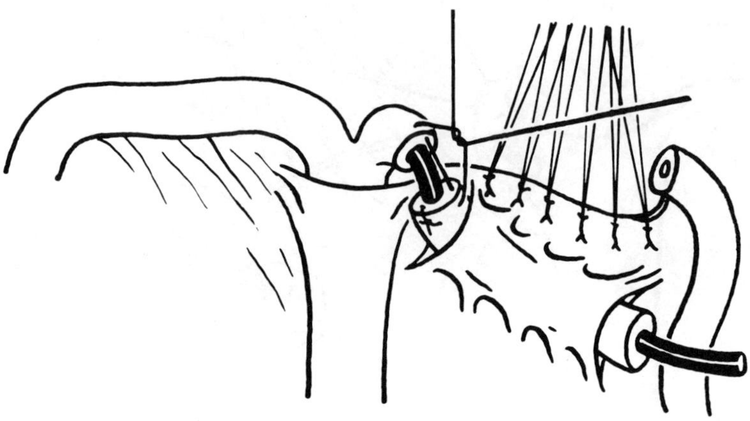

Fig. 4a

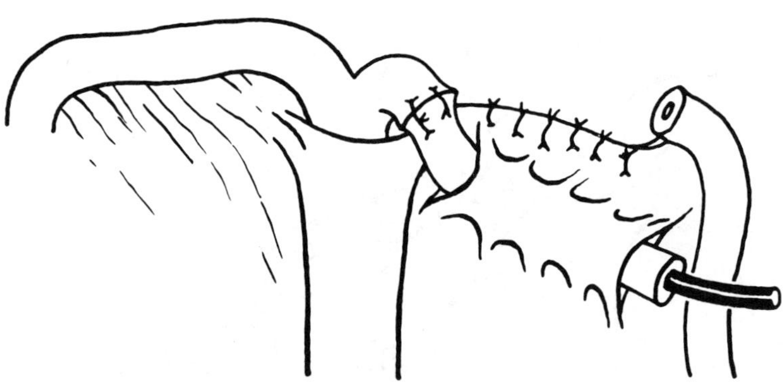

Fig. 4b

- 7 times hydrometra or pyometra were detected in the peripheral portion of the transplant
- 7 times the transplant was incorporated at least in the central suture.

The results of homotransplants thus are not as encouraging /satisfactory / as after autotransplantations. We expect better results after the use of immunosuppressive substances and after operations performed by a modified surgical technique which will render a more satisfactory nutrition of the transplant possible and will approach the procedure which in future could be used in humans.

In five pairs of female rabbits the transplant slipped onto the polyethylene catheter was passed through a tunnel formed beneath the serous membrane of the plica lata / figs. 1,2/. Ten times the transplant was stitched only centrally, the peripheral end was left loose projecting into the abdominal cavity / 4a, 4b /. The catheter which reached 1 cm beyond the peripheral pole of the transplant was fixed to the serous membrane of the plica lata.

Four to six weeks after operation the animals were subjected to relaporotomy and the catheter was removed. Part of the animals are being followed up. We are proceeding with our experiments as the hitherto assembled results promise that this modification will be more suitable and more useful.

THE USE OF BLASTOCYSTS IN HUMAN AND VETERINARY MEDICINE

DISCUSSION ON THE USE OF BLASTOCYSTS IN HUMAN AND VETERINARY MEDICINE

R. G. Edwards

Physiological Laboratory

Cambridge University, U. K.

I wish to raise three aspects of our work on blastocysts. The first concerns the expression of transplantation antigens on mouse blastocysts, a topic of considerable interest in studies on immunoreproduction. The presence of these antigens on mouse embryos at these stages has been reported by Simmons & Russell (1966) and Olds (1968), the former using ectopic transfers of blastocysts to immunised and normal recipients as a test system, the latter using mixed antiglobulin tests on the embryos. We have also used mixed antiglobulin tests (Sell, Coombs & Edwards, unpublished), and employed antisera from mice specifically immunised with spleen cells.

In no case could we find a genuine positive reaction between morulae or blastocysts and antiserum in tests for the histocompatibility antigens. Under certain conditions a false positive could be obtained. If the embryos were left in the antiserum for longer than an hour, or if carrier protein was not added to the medium, non-specific adsorption of antibody to the embryo could be detected without difficulty. Under these circumstances, various types of antibody could be persuaded to adhere to the embryos. When care was taken to avoid these conditions, all tests were negative. We therefore conclude that mouse trophoblast does not possess detectable amounts of the H2 antigens.

The trophoblast could thus provide a protective barrier against immunological reactions by the mother during early pregnancy, and perhaps during later stages too. Larger amounts of histocompatibility antigens differentiate in foetuses during middle and late pregnancy (Edidin, 1964). Other protective measures are evidently invoked to protect the foetus at these stages. One of these is evidently enhancement facilitation specific for the transplantation antigens of the foetus (Hellström, Hellström & Brawn, 1969). This and other forms of protection, e.g. tolerance, might develop in mid-pregnancy when mouse females become less able to reject grafts (Prehn, 1960; Bryere & Barrett, 1961; Currie, 1969; Volkova & Maysky, 1969).

My second topic concerns the manipulation of the rabbit and sheep blastocyst, especially with reference to 'typing' blastocysts for various genetic characteristics. Fragments of trophoblast can be excised from rabbit and sheep blastocysts (Rowson & Moore, 1966; Gardner & Edwards, 1968), and many of the embryos are capable of repair and development to full term. The excised fragment can be used to type the embryo, e.g. for sex, for the sex chromatin body could be identified in cells of the fragment excised from rabbit blastocysts. By transferring sexed blastocysts to recipient females, complete control over the sex ratio at full term was achieved (Gardner & Edwards, 1968). Recently, the quinacrine dyes have been used to detect the Y chromosome in interphase or mitotic cells (Pearson, Bobrow & Vosa, 1970; George, 1970). These stains should assist greatly in determining the sex of blastocysts.

My third topic concerns the more remarkable manipulations on mouse blastocysts carried out by Gardner (1968, 1970). These blastocysts can be literally turned inside out, their inner cell mass removed and replaced by another, the embryo reconstituted, and then develop to full term in a recipient female. Two or more inner cell masses can be placed in one blastocyst, and many of the foetuses then develop as chimaeras. Groups of 1 - 3 cells can be placed into the blastocoel of a recipient blastocyst, and will then colonise large areas of the foetus during development.

This type of work opens the possibility of unique scientific and medical work. Chimaeras could be produced with particular organs colonised by donor tissue. Scientifically, this could be an elegant test of cellular differentiation; medically, it could be a potential tool in the alleviation of genetic disorders although no-one has yet begun to debate the medical opportunities. Major areas of differentiation have already occurred in very early embryos, a few cells being the sole primordia of major organ systems (e.g. Gartler, 1970). Thus, placing a few cells of known potentiality into a blastocyst could impose major changes on foetal development.

ACKNOWLEDGMENT

I am indebted to Mr R. L. Gardner for permission to quote his data on mouse blastocysts.

REFERENCES

Bryere, E. J. & Barrett, M. K. 1961. J. nat. Cancer Inst. 27, 409.
Currie, G. A. 1969. In Foetal Autonomy. Ciba Foundation Symposium. J. & A. Churchill, London.
Edidin, M. 1964. Transplantation, 2, 627.
Gardner, R. L. 1968. Nature, Lond. 220, 596.
Gardner, R. L. 1970. In Intrinsic and Extrinsic Factors in Early Mammalian Development. Ed. G. Raspé. Advances in the Biosciences, 6 (in press). Pergamon: Vieweg.
Gardner, R. L. & Edwards, R. G. 1968. Nature, Lond. 218, 346.
Gartler, S. M. 1970. In Intrinsic and Extrinsic Factors in Early Mammalian Development. Ed. G. Raspé. Advances in the Biosciences, 6 (in press). Pergamon: Vieweg.
George, K. P. Nature, Lond. 226, 80, 1970.
Hellström, K. E., Hellström, I. & Brawn, J. 1969. Nature, Lond. 224, 914.
Olds, P. J. 1968. Transplantation, 6, 478.
Pearson, P. L., Bobrow, M., & Vosa, C. G. 1970. Nature, Lond. 226, 78.
Prehn, R. T. 1960. J. nat. Cancer Inst. 25, 883.
Rowson, L. E. A. & Moore, R. M. 1966. J. Anat. 100, 77.
Simmons, R. L. & Russell, P. S. 1966. Ann. N. Y. Acad. Sci. 129, 35.
Volkova, L. S. & Maysky, I. N. 1969 In Immunology & Reproduction. Ed. R. G. Edwards. International Planned Parenthood Federation, London.

HOW DISTANT IS THE CLINICAL USE OF HUMAN BLASTOCYSTS?

R. G. Edwards

Physiological Laboratory

Cambridge University, U. K.

The kind invitation to speak at this conference allows me to describe our work on human conception from my viewpoint as a scientist. Patrick Steptoe, my medical collaborator, is in the hall and I am sure that he will be delighted to deal with the medical viewpoint. I will concentrate on our own work today, and not make a detailed review of the contributions of others.

The purposes of our work have been to obtain knowledge on fundamental aspects of human conception and to develop methods for studying early human embryos for scientific and clinical purposes. Various problems have arisen, both scientific and clinical. An early difficulty was to obtain a supply of human oocytes that were ready for fertilization, for ovarian oocytes are in the dictyate stage and not ready for fertilization.

The numbers of ova that can be taken from pieces of whole excised ovaries are considerable, abd would obviously provide an excellent source of material. We were able to utilise this supply of oocytes by inducing the final stages of maturation in culture. This was achieved simply by removing oocytes from their follicles and placing them in a suitable culture medium. After several hours in culture, oocyte maturation began as shown by the resumption of meiosis at diakinesis, to be followed by metaphase, anaphase and telophase of the first meiotic division, the extrusion of the first polar body, and after approximately 36 hr in culture, development to metaphase of the second meiotic division (metaphase-II). For culture, various artificial media were supplemented with calf or human sera, and the oocytes developed synchronously and predictably through these stages of meiosis (Edwards, 1965). When different media were used for culture, especially those with high levels of pyruvate, seemed to hasten maturation.

When we were confident that oocyte maturation was proceeding, we could begin the study of three aspects of early human development:

1. Since animal oocytes mature <u>in vitro</u> at the same rate as they do after LH stimulation <u>in vivo</u>, we could infer that in man oocytes would be in metaphase-II 36 hr after endogenous LH release or the injection of exogenous HCG. In other words, the interval between the LH surge and ovulation would be 36 hr.

2. With the timings of diakinesis and later meiotic stages under control, we could begin to study the chromosomal events in meiosis leading to Mongolism and other forms of trisomy.

3. When oocytes reached metaphase-II they were ready for fertili ation.

I will describe the chromosomal events very briefly, and concentrate mostly on studies on fertili ation and cleavage.

THE ORIGIN OF HUMAN TRISOMY

If we are to use human blastocysts clinically, it is essential to understand the origin of human chromosomal trisomics, in order to avert if possible the birth of afflicted children. Various hypotheses exist to explain the origin of trisomy (see Edwards & Fowler, 1970). We have based our studies on the detailed analysis of mouse chromosomes (Henderson & Edwards, 1968). In this species the pairing of homologous chromosomes in diakinesis became progressively weaker as oocytes were taken from older mothers. Many oocytes taken from old mice were found to possess chromosomes that were separated completely. These conditions are likely to result in abnormal chromosome segregation into the ovum and first polar body. Human chromosomes in oocytes show similar configurations (Edwards & Henderson, unpublished; Edwards, 1970). In contrast, examination of chromosomes in the testis reveals that pairing is consant, except occasionally for the XY bivalent (Kjessler, 1970).

These observations lead to the tentative conclusion that autosomal trisomics arise in the oocyte, whereas sex chromosome trisomy can arise in either oocyte or spermatocyte. There is confirmatory evidence for this suggestion. Using the sex-linked marker Xg^a, Race & Sanger (1969) have shown that sex-chromosome trisomics can arise in either parent. In contrast, the frequency of many autosomal trisomics rises with increasing maternal age (Polani, 1969), which obviously implicates the oocyte.

The question needing answer is why changes occur in oocytes as the mother grows older. Consideration of various genetic and cytogenetic data leads to an unexpected conclusion. It appears that the oocytes in a female are produced in a regular order in the foetus, and are then used in the same order in adult life. But those eggs formed later in the foetal ovary are produced under poor physiological conditions, which leads to abnormal pairing and association of chromosomes (Henderson & Edwards, 1968). Eggs formed later are then ovulated late in life, leading to a correlation between the incidence of trisomy and advancing maternal age. Our suggestion of a 'production line' of eggs of this nature is supported by various observations on the formation of oocytes in the foetal ovary, for a definite sequence can be seen histologically (e.g. van Wagenen & Simpson, 1965; Edwards & Fowler, 1970).

FERTILIZATION OF HUMAN EGGS IN VITRO

For several years, the fertilization of human eggs *in vitro* was difficult to achieve, and the incidence of successful fertilization was low (Edwards et al, 1966). But a new line of investigation arose from observations on hamster eggs which showed that fertilization *in vitro* was possible using epididymal spermatozoa (Yanagimachi & Chang, 1964; Barros & Austin, 1967). The necessary conditions in the culture medium were defined by Bavister (1969), who showed that a high pH was essential. Adoption of this system to human fertilization proved immediately successful, a large proportion of eggs being penetrated (Edwards, Bavister & Steptoe, 1969; Bavister, Edwards & Steptoe, 1969). The timings of human fertilization are shown in Table 1.

TABLE 1. TIMINGS OF HUMAN FERTILIZATION

Hours after insemination	Unpenetrated	Sperm in peri-vitelline space	Sperm head in vitellus	Pro-nucleate
6-6½	3			
7-10½	5	4		
11 and later	12	5	1	11

So far, we had been using oocytes matured _in vitro_ to metaphase-II. In animals, these oocytes are incapable of giving rise to full-term foetuses after fertilization, for degeneration occurs soon after fertilization (Chang, 1955; Edwards, unpublished). It was now necessary to obtain oocytes from women shortly before ovulation, and to use the oocytes for fertilization. In animals, pre-ovulatory oocytes recovered from their follicles can develop into full-term foetuses.

RECOVERY OF PRE-OVULATORY OOCYTES FROM WOMEN TREATED WITH GONADOTROPHINS

It thus became necessary to take control of follicular maturation and induce the changes leading to ovulation. From our previous studies described above we had predicted that diakinesis would occur 28 - 30 hr, and ovulation 36 hr after the injection of LH. Patients were therefore given priming doses of human postmenopausal gonadotrophin (Pergonal, G.D.Searle & Co.), starting with low dosages and working to higher doses with increasing experience. We had previous observations on similar treatment of acyclic women by Gemzell, Lunenfeld and Crooke to guide us, and we were aware of factors such as the synergism between FSH and LH (Crooke, 1970) and the possibility of inducing the hyperstimulation syndrome (Mozes et al, 1965). Total doses of a maximum of nine vials of Pergonal given in three equal injections on days 3, 5, and 8 (with some slight variation) were finally adopted, followed by an ovulatory dose of 5,000 i.u. of human chorionic gonadotrophin (Pregnyl, Organon) on day 10 or 11. Laparoscopy was used to aspirate the follicles and recover the oocytes, and was carried out between 28 and 32 hours after the injection of HCG (Steptoe, 1967; Steptoe & Edwards, 1970).

Many preovulatory oocytes could be recovered from the patients (Steptoe & Edwards, 1970). A small suction device was constructed for aspiration (Purdy, Sekker, Edwards & Steptoe, unpublished), and with it single follicles could be cleanly aspirated. Follicular response to the treatment was good (Table 2), although there was no significant differences in the number of follicles found after different doses of Pergonal.

The ovulatory follicles could be identified from the stages of meiosis identified in the oocytes. Flattened preparations of oocytes permitted the distinction between preovulatory and non-ovulatory oocytes to be made; the former were in diakinesis or metaphase-I, the latter were in dictyotene. Follicular fluids were also collected, and the analysis of their steroidal content has begun (Abraham, Odell, Edwards & Purdy, 1970; and unpublished). We want to know if differences exist in the steroids of ovulatory and non-ovulatory follicles. Approximately 35%of follicles yielded an oocyte.

TABLE 2. FOLLICULAR RESPONSE TO TREATMENT WITH GONADOTROPHINS

Dose of HMG (i.u.)	Mean number of follicles
150	4.4 ± 0.7
300 - 375	4.6 ± 0.3
675	5.6 ± 0.4

We had to ensure next that the cycle induced in the patients was suitable for establishing a pregnancy should embryos be placed in their uterus. Three criteris indicated that conditions were satisfactory in many of the patients (Steptoe & Edwards, 1970). First, the interval between the injection of HCG and the return to menstruation was fairly regular after the injection of higher doses of hormone. Secondly, endometrial biopsies from some patients were secretory. Lastly, pregnanediol was excreted in large quantities a few days after aspiration of the follicles by almost all of the patients examined. These criteria thus indicate that a suitable uterine response could be expected to the treatment with gonadotrophins.

CULTURING PRE-IMPLANTATION HUMAN EMBRYOS

Preovulatory oocytes could thus be taken directly from the patients, and the conditions for fertilization _in vitro_ had been identified. We could therefore begin to study the cleavage of human eggs in culture, using oocytes from the wives and spermatozoa from the husbands. A medium developed by Whittingham (1968) permitted the growth of mouse ova through fertilization and the first cleavage _in vitro_, and we selected this medium as a basis for studies on human embryos. Human eggs cleaved to the 8-celled stage in this medium, under a gas phase of 5% CO_2 in air, but not beyond. Cleavage is regular, and the embryos appear to be in excellent condition (Figure 1). We are now testing other media to find out if development beyond the 8-celled stage can be secured.

Our next objectives will be to obtain further embryonic development _in vitro_, and to devise methods for the transfer of embryos into the human uterus. Two or three more cleavages should result in blastocysts. Development to the 8-celled stage is probably sufficient to permit the embryos to develop further when placed in the uterus. Further growth in culture might be simply

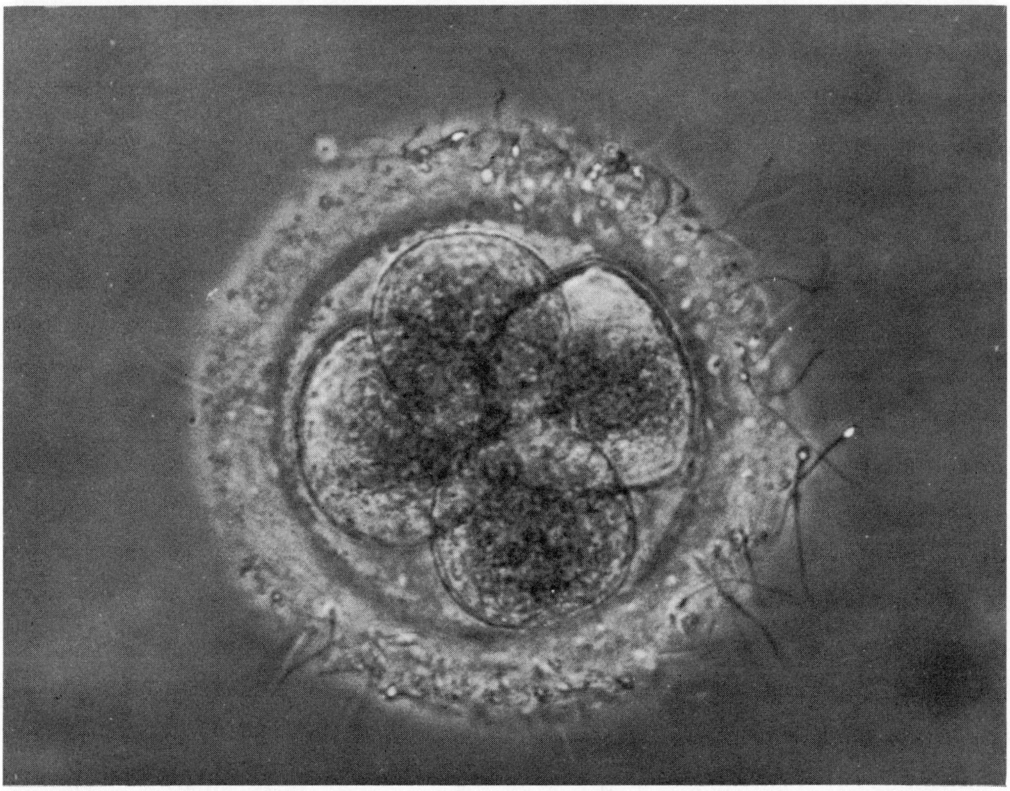

Figure 1. A 4-celled human egg fertilised and cultured in vitro. It was slightly compressed for photography.

a matter of the correct gas phase, for Whitten (1970) has shown that a reduced oxygen tension permits all 1-celled mouse eggs in culture to develop to blastocysts in a medium not greatly dissimilar from Whittingham's medium. Alternatively, the rabbit might serve as a host for human embryos, as it does for those of other species (Averill, Adams & Rowson, 1958). Should none of these methods prove suitable, it might be necessary to expose the embryos to uterine secretions. Such secretions appear to be important for embryonic development in some species, e.g. the rabbit (Beier, 1970; Daniel, 1970). We have constructed a small chamber which can be inserted and removed from human uteri (Edwards et al, 1968), and it might serve to enclose the embryos in a uterine environment, or serve to provide a supply of human uterine fluid.

ACKNOWLEDGMENT

I am deeply indebted to the Ford Foundation for their financial assistance in support of this work.

REFERENCES

Abraham, G.E., Odell, W.D., Edwards, R., & Purdy, J.M. 1970. Second Karolinska Symposium on Research Methods in Reproductive Endocrinology, p. 332. Ed. E. Diczfalusy.

Averill, R.L.W., Adams, C.E., & Rowson, L.E.A. 1958. J. Endocr. 16, 326.

Barros, C. & Austin, C.R. 1967. J. exp. Zool. 166, 317.

Bavister, B.D. 1969 J. Reprod. Fert. 18, 544.

Bavister, B.D., Edwards, R.G., & Steptoe, P.C. 1969 J. Reprod. Fert. 20, 159.

Beier, H.M. In Intrinsic and Extrinsic Factors in Early Mammalian Development. Ed. G. Raspé. Advances in the Biosciences, 6 (in press). Pergamon: Vieweg. 1970.

Chang, M.C. 1955 J. exp. Zool. 128, 379.

Crooke, A.C. 1970 Br. med. Bull. 26, 17.

Daniel, J.C. In Intrinsic and Extrinsic Factors in Early Mammalian Development. Ed. G. Raspé. Advances in the Biosciences, 6 (in press). Pergamon: Vieweg. 1970.

Edwards, R.G. 1965. Lancet, ii, 926.

Edwards, R.G. 1970. Proc. 6th World Cong. Fertility & Sterility, Tel Aviv, p 64. Israel Academy of Sciences & Humanities.

Edwards, R.G., Bavister, B.D., & Steptoe, P.C. 1969. Nature, Lond., 221, 632.

Edwards, R.G., Donahue, R.P., Baramki, T.A., & Jones, H.W. 1966. Amer J. Obstet. Gynec. 96, 192.

Edwards, R.G. & Fowler, R.E. 1970. Modern Trends in Human Genetics, p 181. Ed A.E.H. Emery. Butterworths.

Edwards, R.G., Talbert, L., Israelstam, D., Nino, H.V. & Johnson, M.H. 1968. Amer. J. Obstet. Gynec. 102, 388.

Henderson, S.A. & Edwards, R.G. 1968 Nature, Lond., 218, 22.

Kjessler, B. 1970 Modern Trends in Human Genetics, p. 214. Ed. A.E.H. Emery. Butterworths.

Mozes, M., Bogokowski, H., Antebi, E., Lunenfeld, B., Rabau, E., Serr, D.M., David, A., & Salomy, M. 1965 Lancet, ii, 1213.

Polani, P.E. 1969 Br. med. Bull. 25, 81.

Race, R.R. & Sanger, R. 1969 Br. med. Bull. 25, 99.

Steptoe, P.C. 1967. Laparoscopy in Gynaecology. E. & S. Livingstone, Edinburgh.

Steptoe, P.C. & Edwards, R.G. 1970 Lancet, i, 683.

Whitten, W.K. 1970. In Intrinsic and Extrinsic Factors in Early Mammalian Development. Ed. G. Raspé. Advances in the Biosciences, 6 (in press). Pergamon: Vieweg.

Whittingham, D.G. 1968. Nature, Lond., 220, 592.

Yanagimachi, R. & Chang, M.C. 1964 J. exp. Zool. 158, 361.

LAPAROSCOPIC RECOVERY OF PRE-OVULATORY HUMAN OOCYTES AFTER PRIMING OF THE OVARIES WITH GONADOTROPHINS

P. C. Steptoe, F.R.C.S. (Ed.), F.R.C.O.G.

Oldham and District General Hospital

Lancashire, England

Human oocytes matured and fertilized in vitro (Edwards, 1965; Edwards, Bavister, and Steptoe, 1969; Bavister, Edwards, and Steptoe, 1969) are now available for laboratory studies, but two outstanding problems must be solved before they can be used clinically. First, pre-ovulatory oocytes must be recovered after they have completed much of their maturation in the ovary.

Secondly, methods must be devised for the recovery of oocytes from the human ovary. Laparoscopy should be invaluable for this purpose: it makes minimum demand on the patient, permits many manipulations in the abdominal cavity, and can be used repeatedly in the same patient (Steptoe, 1969).

Human postmenopausal and chorionic gonadotrphins (HMG and HCG) were therefore used in cyclic women to impose some control over the menstrual cycle and oocyte maturation, and laparoscopy was performed to aspirate follicles shortly before the expected time of ovulation. Follicular development must be timed accurately, and our guide-lines were based on the timing of oocyte maturation in vitro (Edwards, 1965); in animals the durations of oocyte maturation in vitro and in vivo have proved to be very similar. In culture, human oocytes reached metaphase of the first meiotic division (metaphase-1) after 28 to 32 hours, and metaphase-11 after 36 hours; human ovulation should therefore occur 36 hours after the ovulatory injection of HCG. Some evidence already indicates that these estimates correctly measure the timing of maturation in vivo (Jagiello, Karnicki, and Ryan, 1968). Laparoscopy was therefore performed 29 to 31 hours after the injection of HCG, for diakinesis or metaphase-1 are easily recognised and provide an excellent marker of the exact stage of maturation. Variations might exist between women in the timing of oocyte maturation, although in animals such variation is very small.

The Patients

The volunteer patients were the women of infertile married couples numbering 46 in all. All couples were subjected to full investigation, including laparoscopic examination of the wife. The patients selected for oocyte recovery were those with peritubal pathology, tubal occlusion, absence of the tubes, failed tubo plastic surgery and male oligospermia. The object of the investigations was fully discussed with the patients, including the possible clinical applications to relieve their infertility. In some of the patients pelvic adhesions created difficulties for the examination of the ovaries and the aspiration of follicles.

MATERIAL AND METHODS

Treatment with Gonadotrophins

Most of the patients were menstruating regularly. Low doses of HMG were used initially since very little information exists on the treatment of cyclic women with gonadotrophins. The patients would be secreting endogenous FSH and LH, and LH greatly enhances the response to FSH in amenorrhoeic or anovulatory patients (Crooke, 1970). Moreover, higher doses of gonadotrophins more commonly induce the hyperstimulation syndrome (Mozes, Bogokowsky, Antebi, Lunenfeld, Rabau, Serr, David, and Salomy, 1965).

Initially two vials of HMG (75 i.u. per vial) were given intramuscularly in one injection; this dosage was gradually increased to three injections of three vials each. Treatment began early in the menstrual cycle, although not always on the same day, because of the limited times when laparoscopy could be performed. Patients given one injection were treated between days three and five of the menstrual cycle, those given two injections between days three and seven, and those given three injections between days two and nine. The interval between successive injections was either two or three days. To induce ovulation, one injection of HCG, containing between 1,000 i.u., and 12,500 i.u., was given intramuscularly between days nine and eleven, preferably on day ten. The patients were brought into hospital on the day after the injection of HCG for the aspiration of Graafian follicles by laparoscopy.

Examination of Follicular Aspirates

The colour and amount of fluid from each follicle was first assessed macroscopically, and the fluid was then searched microscopically for oocytes. Each fluid sample was centrifuged to deposit cellular matter, and frozen for storage. Delays of up to three hours could intervene between collection and freezing of these samples.

Examination of the Oocytes

Oocytes were classified by the presence or absence of cumulus cells. Oocytes with few or no cells were considered to be atretic; those with several layers of cells were classified as "pre-ovulatory" or "non-ovulatory." Pre-ovulatory oocytes were surrounded by cumulus which was diffuse, "silvery" in appearance, slightly sticky and embedded in a viscous matrix.

Some oocytes were immediately fixed in formalin, and the remainder were made into flattened preparations. Most or all of the cumulus and corona cells were removed during pretreatment with hypotonic sodium citrate (1.5%) for 20-30 minutes, and the oocytes were stained with aceto-orcein. Prometaphase, diakinesis, metaphase-1, or (rarely) metaphase-11 indicated that the oocyte was pre-ovulatory, whereas non-ovulatory oocytes possessed a germinal vesicle. Atretic oocytes were in various stages, including metaphase-11.

RESULTS

A total of 59 courses of treatment were given to 46 patients. Full data are not available from all of the courses, for evaluation of the early results led to fuller observations being made later.

Every patient possessed Graafian follicles, and only one of them displayed an apparently fresh corpus luteum. Ovulation had therefore not occurred in 58 of the patients, although it was imminent in many of them as judged from the morphology of the follicles.

The presence of many follicles was responsible for somewhat enlarged ovaries in some patients. Ovarian volume never exceeded one and half times normal size. The largest follicle was 4 cm. in diameter, and no enlarged cysts were seen.

Recovery of Oocytes from Follicles

Detailed results came from 49 patients. Of the 368 aspirated follicles, 118 yielded oocytes, i.e. 32% (Table 2). Two or more oocytes were occasionally aspirated from one follicle, and 133 oocytes were recovered altogether. Some of the aspirations were poor because new methods were being tested, the follicle was not fully penetrated, or the opposite wall of the follicle was damaged.

Size of Follicle in Relation to Recovery of an Oocyte

Follicular size was estimated through the laparoscope, and is related to the recovery of oocytes. Recoveries were highest from follicles between 1 cm. and 2 cm. in size. The lower recoveries from larger follicles might have been due to the location of oocytes near the stigma, or to the loss of fluid, especially when the thin wall was penetrated.

Types of Oocyte Recovered

A total of 97 oocytes was classified microscopically. 12 were atretic and 43 non-ovulatory. 42 were evidently pre-ovulatory. 18 were in metaphase-1, 3 in metaphase-11 and 12 in diakinesis or prometaphase at recovery; 9 others reached metaphase-11 after culture.

Pre-ovulatory oocytes were more commonly aspirated from larger follicles, although some of them came from follicles of ½ cm. to 1 cm. in size. During later stages of the work, patients given 675 i.u. HMG yielded an average of almost one pre-ovulatory oocyte per ovary.

DISCUSSION

The Use of Aspirated Oocytes in Establishing Pregnancy

Laparoscopy may be used to place oocytes and embryos in the oviduct or uterus, although our initial attempts to transfer oocytes to the oviduct have been only partially successful (Steptoe and Edwards, unpublished). Alternatively, entry into the abdominal cavity might be avoided by transferring the embryos into the uterus via the cervix, a method yielding some success in cattle (Rowson and Moor, 1966). Knowledge is now needed about the viability of human oocytes and fertilized eggs. The oocytes are recovered during metaphase-1, which could be a sensitive stage for the induction of trisomy or polysomy. But until morulae or blastocysts can be examined, it will be difficult to obtain sufficient mitoses to determine the chromosomal complement of the embryo. Chromosomal or other anomalies could also arise during cleavage through adverse conditions in the media.

The control of the pre-implantation development of human embryos can now be attempted with improved chances of success. Should difficulties arise, the rabbit oviduct might nourish human eggs, especially from the two-celled stage, as it does those of other species (Averill, Adams, and Rowson, 1958). Unfertilized human oocytes have already been observed to remain in excellent condition in the rabbit oviduct (Edwards, Donahue, Baramki, and Jones, 1966).

REFERENCES

Averill, R. L. W., Adams, C. E., and Rowson, L. E. A. (1958); J. Endocr. 16, 326.

Bavister, B. D., Edwards, R. G., and Steptoe, P. C. (1969); J. Reprod. Fert. 20, 159.

Crooke, A. C. (1970); Br. Med. Bull. 26, 17.

Edwards, R. G. (1965); Lancet. ii, 926.

Edwards, R. G., Bavister, B. D., and Steptoe, P. C. (1969); Nature, Lond. 221, 632.

Edwards, R. G., Donahue, R. P., Baramki, T. A., and Jones, H. W. (1966); Am. J. Obstet. Gynaec. 96, 192.

Jagiello, G., Karnicki, J., and Ryan, R. J. (1968); Lancet. i, 178.

Mozes, M., Bogokowski, H., Antebi, E., Lunenfeld, B., Rabau, E., Serr, D. M., David, A., and Salomy, M. (1965); Lancet. (Dec.) ii, 1213-1215.

Rowson, L. E. A. and Moor, R. M. (1966); J. Reprod. Fert. 11, 311.

Steptoe, P. C. (1969); Laparoscopy: Diagnostic and Therapeutic Uses. In Proc. of The Royal Soc. of Med. 62, 439-441.

SCANNING ELECTRON MICROSCOPY OF MOUSE BLASTOCYSTS BEFORE AND AT IMPLANTATION

Bergström,S., and Nilsson,O.

Institute of Human Anatomy, S752 20 Uppsala, Sweden

Mouse blastocysts spend the last few days before implantation in the uterine cavity. They arrive there coated by the zona pellucida, which a day later has disappeared. After this zona-loss the blastocysts attach to the uterine epithelium and implant.

The state with a zona-free blastocyst lying unattached in the uterine lumen is normally short, at most a few hours. This time can be prolonged experimentally by inducing a so-called delayed implantation. That is, if early pregnant mice are spayed and given progesterone daily the blastocysts mature and shed their zonae but the further development is stopped and the blastocysts become dormant. This state remains until estrogen is injected in addition to progesterone. Then the blastocyst is activated, it implants, and the pregnancy continues. If this technique is used, some stages of blastocyst development thus can be made accessible for detailed study in a convenient way.

The blastocysts of the present experiments were obtained during experimentally delayed implantation. They were rinsed out of the uterus with a phosphate-buffered solution of glutaraldehyde, post-fixed in OsO_4 and then prepared by freeze-drying for scanning electron microscopy.

The various functional states of the blastocyst are shown in Table 1. First being coated by the zona the blastocyst is incapable to implant: it is <u>inert</u>. After <u>divesting</u> itself of the zona it reaches the state that can be prolonged by experimentally delayed implantation. This is the dormant state of the blastocyst. When we terminate this state by injecting estrogen the blastocyst gets <u>activated</u> and attachment is subsequently induced.

FUNCTIONAL STATES OF MOUSE BLASTOCYSTS IN THE UTERUS BEFORE AND AT IMPLANTATION

Progestation		
	Zona present:	Inert blastocyst
	Zona disappears:	Divesting "
	Zona absent:	Dormant "
Gestation		
	Attachment induced:	Activated "
	Attachment occurs:	Attaching "

In our current experiments on the blastocyst development we have been looking for a convenient indicator that reveals the functional state of the blastocyst. We have found that the surface of the blastocyst - as seen in the scanning electron microscope - has a characteristic appearance during the various functional states.

The inert blastocyst is characterized by an outer surface of an amorphous kind, the zona pellucida. Early after the entrance in the uterus the blastocyst carries a smooth zona. Some hours later the zona loses its smooth appearance, and a lot of small pores occurs (Fig.1). Gradually the blastocyst is then divesting itself of the zona. The factors causing this shedding process are not known in any detail, and the zona-divesting capacity has been attributed both to the trophoblast cells and to the uterine secretion.

During early stages of delayed implantation a crack in the zona can sometimes be observed. The zona-coated blastocyst is "hatching" but it can be argued whether or not this is an artifact. Beneath the zona, however, the surface of some trophoblast cells is visible. It is a cobble-stone surface, each bulging part being the outer surface of one trophoblast cell. Furthermore, the trophoblast cells have several finger-like microville (Fig.2).

When the whole zona has disappeared the blastocysts can be kept in the dormant state by using the technique of experimentally delayed implantation. Also in this stage we still notive the bulging trophoblast cells and the microvilli.

Induction of implantation can be achieved by giving estrogen in addition to progesterone in an experimentally delayed implantation. The donor of the blastocyst seen in Fig.3 was given estrogen half a day before intraluminal instillation of the fixative. The blastocyst is thus just prior to attachment and is classified as activated. The bulging trophoblast cells and the outline of their cell borders are seen. The microvillous projections are numerous (Figs.4 and 5). We have not yet found, however, any characteristic differences in surface appearance between these activated blastocysts and the dormant ones.

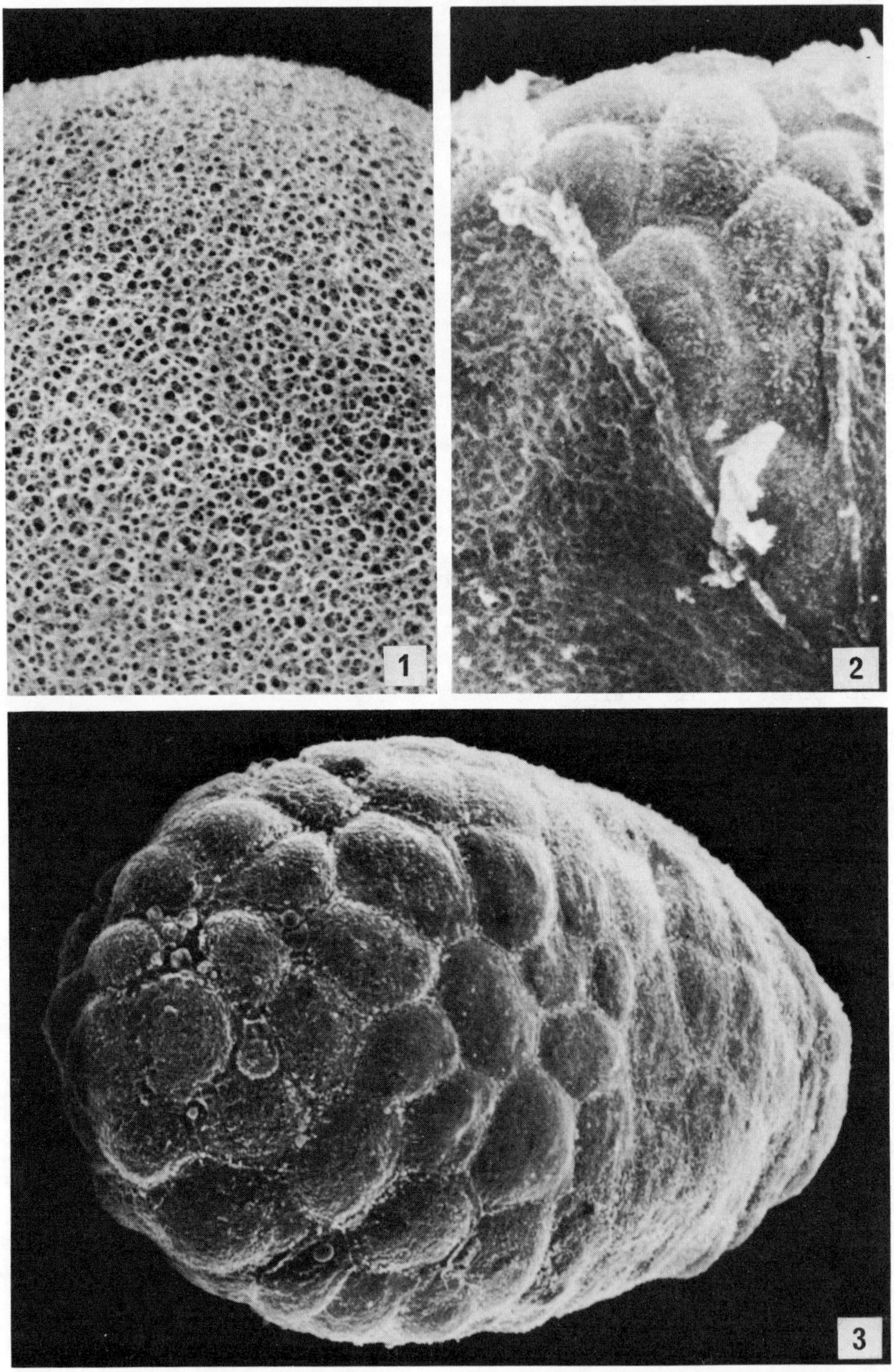
1
2
3

4

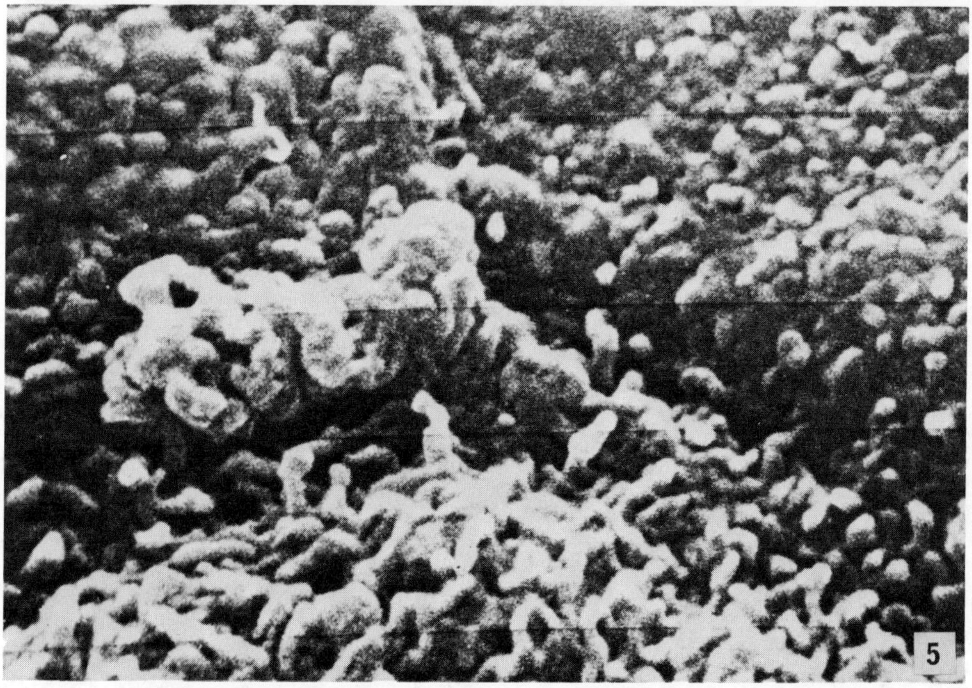
5

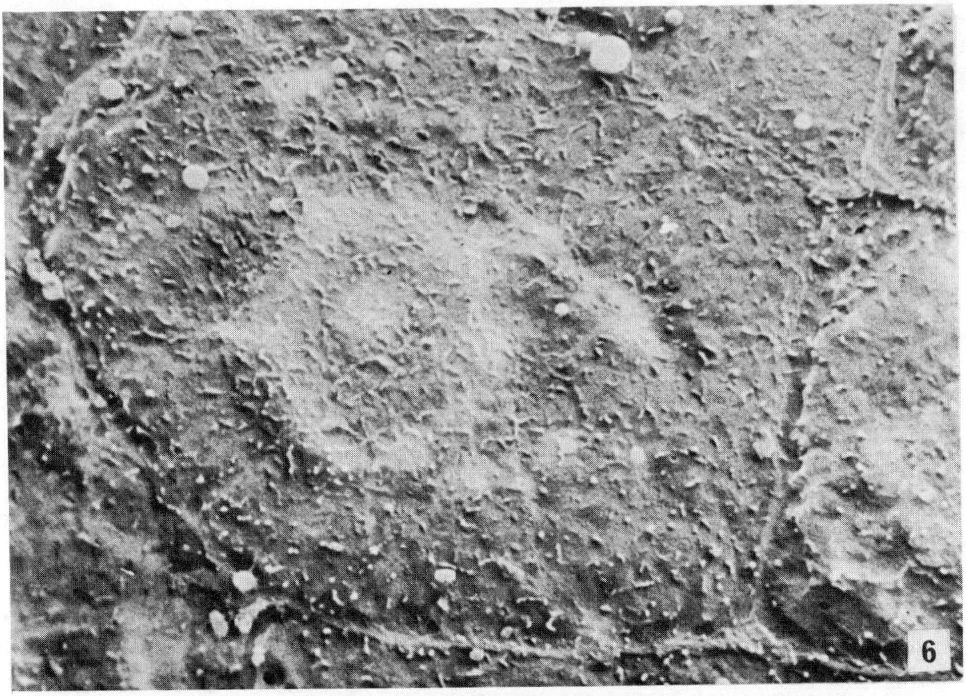

6

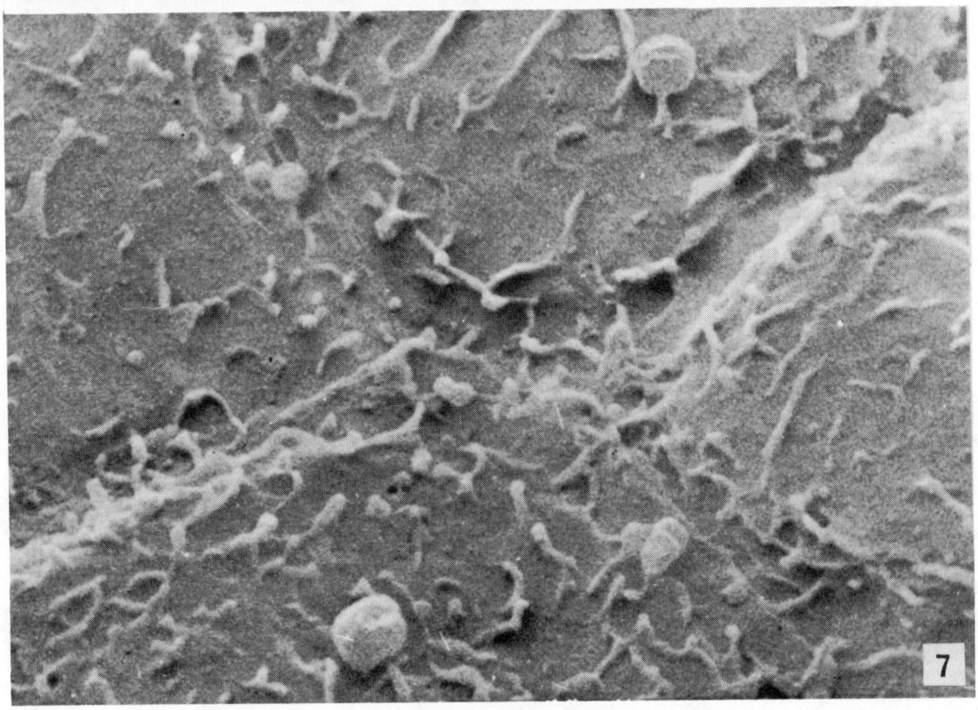

7

Blastocysts obtained about a day after estrogen injection are attaching. Now, the surface of the trophoblast cells is characteristically flattened, the microvilli are few and have the shape of ridge-like processes (Figs. 6 and 7).

To conclude, these preliminary findings indicate that the surface structure of the blastocyst changes with its functional state. We are continuing our experiments and will later report the changes in detail.

Figure Legends

Fig. 1. Inert blastocyst with zona pellucida. Day 4, 4 p.m. of normal pregnancy. Mag. 1 500 X.

Fig. 2. Divesting blastocyst with an opening in the zona pellucida. Day 5, 3 p.m. of experimentally delayed implantation. Mag. 1 500 X.

Fig. 3. Activated blastocyst, 16 hrs after the injection of estrogen. Mag. 800 X.

Fig. 4. Activated blastocyst, same as in Fig.3. Mag. 3 100 X.

Fig. 5. Activated blastocyst, same as in Fig.3. Mag. 15 900 X.

Fig. 6. Attaching blastocyst, 28 hrs after the injection of estrogen. Mag. 4 500 X

Fig. 7. Attaching blastocyst, same as in Fig. 6. Mag. 9 900 X.

This work was supported by the Swedish Medical Research Council (Grant No B70-12X-70).

RECENT ADVANCES IN CONTRACEPTION

LONG ACTING POLYDIMETHYLSILOXANE IMPLANTS FOR CONTRACEPTION

Benagiano G., Carenza L., Ermini M. and Subrizi A. D.
I[a] Clinica Ostetrica e Ginecologica dell'Università di
Roma, Rome, Italy

Polydimethylsiloxane (PDS) possess the property of being permeable to organic compounds including several steroids. Diffusion *in vitro* from PDS films or capsules is proportional to the area and inversely proportionate to membrane thickness (Kincl et al., 1968). Permeation of various substances including steroid hormones, follows Fick's law (Garrett & Chemburkar, 1968; Kratochvil et al., 1970), assuring therefore a remarkable diffusion constancy when incubated in distilled water (Kincl et al., 1968).

Properties of PDS membranes were first applied to the administration of steroid hormones by Dzuik & Cook (1966), who used PDS implants for synchronizing oestrus in the sheep. Following this observation, Segal & Croxatto (1967) reported that oestradiol embedded in PDS tubings gives a biological effect lasting for several months. The biological effectiveness of steroid hormones administered by subcutaneous injection and by PDS implants, was further studied by Chang and Kincl (1970). They found that in various specific biological tests the increase in efficiency ranged from 2 to 30 times using PDS capsules.

In 1966 Segal suggested that PDS capsules filled with synthetic progestogens could become a simple method for the control of human fertility utilizing the principles of the low level luteal supplementation method of Rudel et al. (1965) and of Martinez-Manautou et al. (1967). Among hormonal contraceptive procedures, this is the only one that requires continuous supplementation of steroid hormones; the method is based on the antifertility effect of small doses of synthetic progestogens and does not suppress ovulation in the majority of cases.

In the present paper we wish to report about the possibilities of applying PDS sustained release preparations to the low progestogen method.

DIFFUSION OF MEGESTROL ACETATE FROM PDS IMPLANTS *IN VIVO*

Implants were prepared by filling 2 cm long polydimethylsiloxane tubings (wall thickness 0.42 mm) with approximately 20 mg of ^{14}C-labelled megestrol acetate (6-methyl-17α-acetoxy-4,6-pregnadiene-3,20-dione). This compound is active as an oral contraceptive at a dose of 500 µg pro die. The capsules were incubated *in vitro* until a constant diffusion was reached. In the present study this amounted to 20 µg per 24 hrs. per capsule.

The implants were then inserted subcutaneously in ten adult female rats. Diffusion from the implants was calculated by measuring the total radioactivity present every day in urine and foeces. Under those *in vivo* conditions, absorption from the implants is not constant. Figure No. 1 shows the pattern of excretion of radioactive metabolites during the six month period of observation. It can be seen that at the end of the experiment, the daily output is only one third of the initial value.

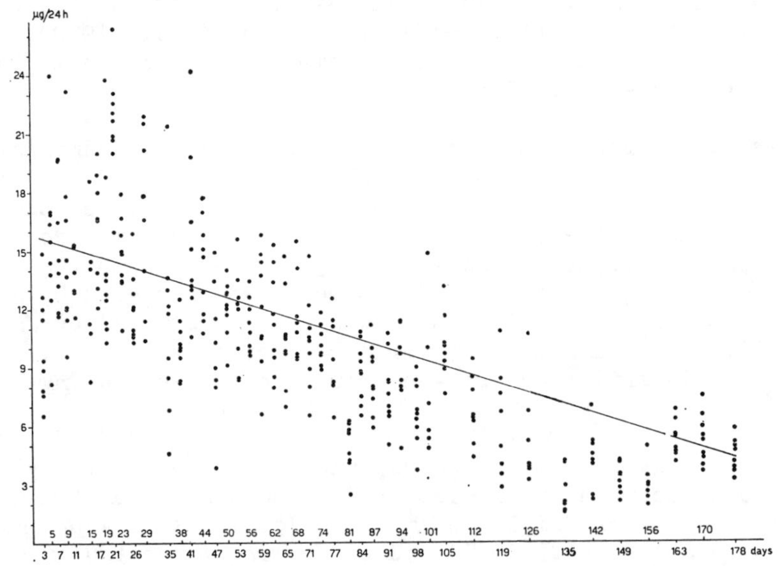

Fig. 1. Total excretion of radioactive metabolites in ten rats inserted with one PDS capsule containing ^{14}C-labelled megestrol acetate (Benagiano & Ermini, 1969).

Due to the results obtained, the same experiment has been repeated in the human using three capsules per person. Ten male healthy volunteers have been selected for this study. Absorption from the implants was calculated on the basis of total urinary metabolites excreted daily. In the human, similar to what observed in the animal, diffusion from the implants is far from being constant: during the first two months decrease is very rapid; however in the subsequent period the absorption tends to stabilize and after six months is approximately 50 % of the initial value.

In order to fully evaluate the importance of local factors interfering with the absorption of the steroid, in 4 cases the capsules were dissected together with the surrounding tissue, at various time interval.

Light microscopy of this tissue demonstrated the presence all around the PDS implants, of a continuous fybrotic wall of variable thickness. A microphotograph of a cross section of one of those pericapsular walls is shown in figure No. 2. It seems likely that formation of fybrotic tissue is one of the factors responsible for the decrease in the absorption in the human.

Other factors however, have to be taken into consideration in evaluating the present results: in some instances PDS capsules have been reincubated in vitro after several months of permanence in the subcute. In this case the amount of megestrol acetate recovered from the desorbing medium was considerably lower than expected. A possible explanation of this finding is the decreased surface contact between crystalline steroid and PDS membrane.

The results obtained so far point out to some limitation of the sustained release preparations: after a certain period of time the efficency of an implant will not be that calculated theoretically, but may decrease to less than 50 %.

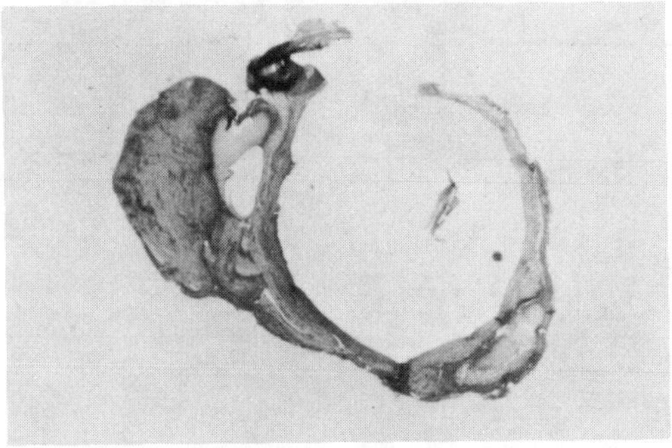

Fig. 2. (x15) Cross section of the newly formed fybrotic tissue surrounding one PDS capsule, six months after the implant in the subcute of the left arm.

EFFECTS OF PDS IMPLANTS ON THE HORMONAL EXCRETION DURING THE MENSTRUAL CYCLE

Preliminary data from other investigators (Croxatto et al.,1969; Tatum et al.,1969) based on several hundred cycles, indicate that implants having the same characteristics of those employed in our experiments are well accepted by the patients, do not give major side effects and are efficent in preventing pregnancies over long periods of time.

For this reason it appeared of interest to study the hormonal excretion pattern in women having PDS implants.

Ten healthy young women aging between 19 and 30, were selected for this research. Their hormonal excretion was followed during one control cycle using daily determinations of urinary FSH and LH, oestrone, oestradiol and oestriol, and pregnanediol. Four capsules releasing each 20 µg pro die in vitro were inserted to 5 of the volunteers, whereas three capsules were implanted in the other 5. The women were then followed, using the same parameters, during the first and the third cycle after the implant.

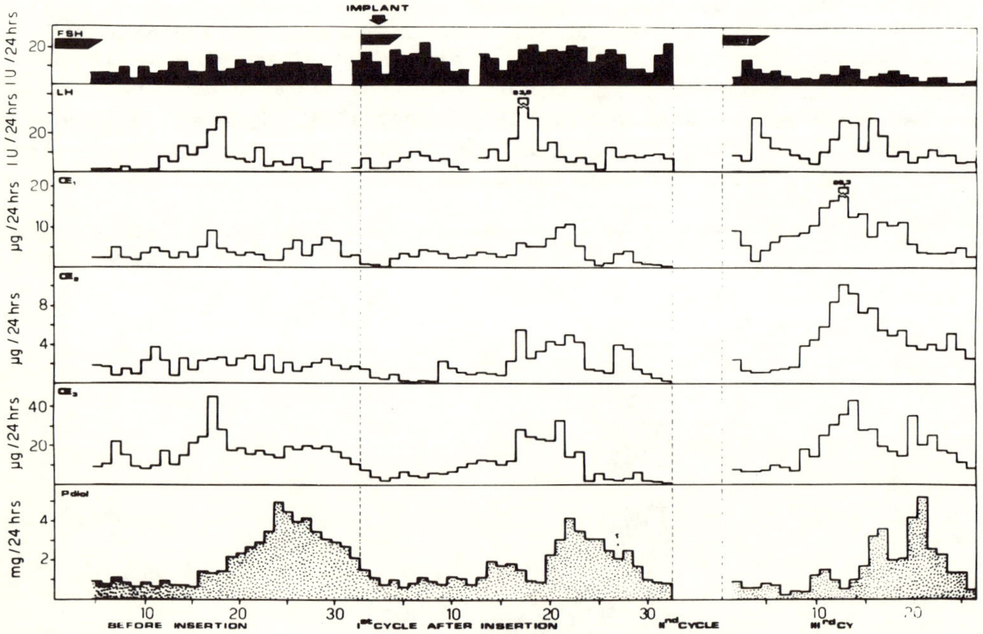

Fig. 3. Urinary hormonal excretion in a 25 year old normally menstruating woman before and after the insertion of four PDS capsules releasing a total of 80 µg of megestrol acetate per 24 hrs., when incubated under in vitro conditions.

The control cycle showed a normal excretion pattern in seven out of ten cases. No major changes occurred in six of them (three with 3 capsules and three with 4) following the insertion of PDS implants: pregnanediol values demonstrated the formation of a functioning corpus luteum, LH excretion followed the ovulatory pattern and oestrogens were eliminated in normal quantities. A representative example of this condition is presented in figure No. 3. On the contrary, significant changes occurred in patient No. 7, as clearly shown in figure No. 4. In this 19 year old girl a certain degree of luteal insufficency was already appearent in the control cycle. Furthermore following the insertion of three capsules the hormonal excretion was completely altered. Clinically she had a menstrual bleeding two weeks after the implant followed by a 27 day cycle and again by two very short (12 and 13 days) anovulatory cycles.

The control cycle was anovulatory in the remaining three patients. The insertion of PDS capsules did not modify substantially this pattern: one cycle showed ovulatory pregnanediol values; all others were anovulatory.

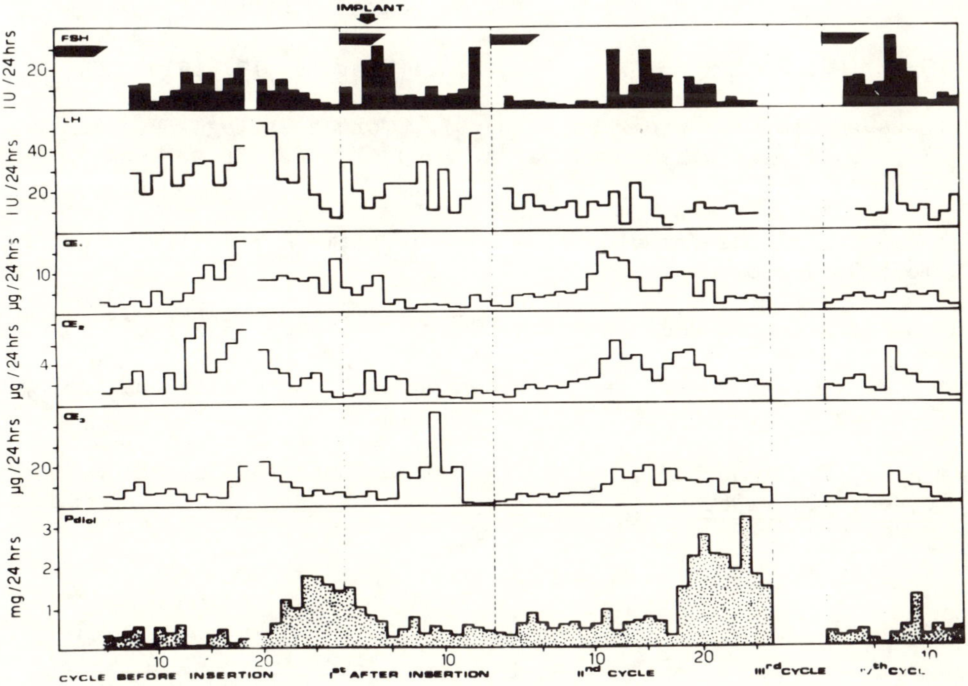

Fig. 4. Urinary hormonal excretion in one case in which substantial modifications were appearent after the insertion of 3 PDS capsules. Pregnanediol values suggest the formation of a corpus luteum only during the second cycle after the implant.

It is obviously difficult to draw conclusions from such small scale experiments; it seems however reasonable to conclude that the antifertility effect of PDS implants filled with megestrol acetate, demonstrated by Croxatto, Tatum and their collaborators is not mediated through an ovulation suppression.

REFERENCES

Benagiano G. & Ermini M.: Clin. Ostet. Ginec. 71 (1969) 51.

Chang C.C. & Kincl F.A.: Fertil. Steril. 21 (1970) 134.

Croxatto H., Diaz S., Vera R., Etchart M. & Atria P.: Am. J. Obst. Gynec. 105 (1969) 1135.

Dzuik P.J. & Cook B.: Endocrinology 78 (1966) 208.

Garrett E.R. & Chemburkar P.P.: J. Pharm. Sci. 57 (1968) 1401.

Kincl F.A., Benagiano G. & Angee I.: Steroids 11 (1968) 673.

Kratochvil P., Benagiano G. & Kincl F.A.: Steroids 15 (1970) 505.

Martinez-Manautou J., Giner-Velasquez J., Cortez-Gallegos V., Aznar R., Rojar B., Gutierrez-Najar A. & Rudel H.W.: Brit. Med. J. 2 (1967) 730.

Rudel H.W., Martinez-Manautou J. & Maqueo-Topete M.: Fertil. Steril. 16 (1965) 158.

Segal S.J.: Excerpta Med. Internat. Congress Series 133 (1966) 1028.

Segal S.J. & Croxatto H.: Paper presented at "23th meeting of the Am. F. Soc." April 14-16 1967, Washington D.C.

Tatum H.J., Coutinho E.M., Filho J.A. & Sant'Anna A.R.S.: Am. J. Obst. Gynec. 105 (1969) 1139.

SUSTAINED RELEASE HORMONAL PREPARATIONS

Fred A. Kincl and Harry W. Rudel

The Population Council, Rockefeller University, N. Y.

and Biological Concepts, Inc., New York, N. Y.

The duration of use of any contraceptive method is predicated on the freedom of side effects, rather than absolute efficacy. This achieved, any method would be acceptable for wide use. The experience obtained with intrauterine devices is illustrative. These have not fulfilled their original expectation because of the problem of side effects. With this method contraceptive effectiveness can be correlated with the size of the device. In order to achieve acceptable efficacy the size of the device must be relatively large which causes uterine distortion, cramps and bleeding and in a sizable percentage of the population the device must be removed.

Because of the high effictiveness, and its acceptability in the western world, hormonal contraception appears to be suitable for developing countries if a long-acting preparation could be obtained.

To achieve long acting preparations of steroid hormones has been the goal of many. Chemical modification of the parent molecule has been the chosen route. Various esters, ethers or enol ethers injected intramuscularly will form a depot providing biological effects lasting several weeks.

Combining cholesterol with steroid hormones was a different approach used. Shimkin and White (1941) mixed various steroids in varying proportions and reported that each hormone and composition was absorbed differently and noted that cholesterol-hormone pellets were absorbed more slowly and more uniformly than pellets of pure hormone. Lipshutz and Iglesias (1961) obtained an anti-fertility effect in mice with pellets containing 20% by weight of 19-norprogesterone and 80% cholesterol that lasted the life span of the animals. Folch Pi *et al* (1965) were able to inhibit fertility in mice by lipid pellets containing only 10% of 19-norprogesterone. Similar results were reported by Kincl et al (1966) in hamsters and rats.

Another approach was used by Dziuk and Cook (1966) who employed silicone rubber (polydimethylsiloxane, PDS) implants containing a progestational agent to synchronize estrus in sheep. They noted that such implants provided an effect lasting several months.

We have studied the feasibility of polydimethylsiloxane capsules and steroid-lipid pellets as sustained release forms. An implant consisting of a polymer membrane permeable to the active substance enclosed therein will maintain the same geometry and surface exposed to body fluids; this may assure a constant absorption as long as the concentration of the active substance within the implant remains sufficiently high. The disadvantages of having to remove steroid exhausted capsules from the implantation site are obvious. It may be therefore preferable to use preparations which would become completely absorbed (such as steroid-lipid pellets), but the disadvantage of such preparations is that the surface area continuously diminishes causing constantly decreasing release of the active principle.

Our studies have shown that in vitro polydimethylsiloxane is permeable to various steroids at rates proportional to the exposed area, and inversely proportional to membrane thickness when the release of the steroid molecules from the membrane into the surrounding desorbing medium is monitored as a function of time.

Polydimethylsiloxane implants for experimental purposes may be made by sealing one end of a PDS tubing of the desired diameter, length and wall thickness with a silastic medical type A adhesive. Dry crystalline material is then packed into the inner cavity and the other end is sealed, maintaining a plug thickness on both ends of approximately 2-3 mm. When incubated in water, after a few days, permeation becomes constant over periods of time up to 40 days. This was shown by measuring accumulation of megestrol acetate released by 1 cm long capsules in 100 ml distilled water. From day 8 to day 22 of incubation the values were 12.5 μg/24 h $\pm$ 0.27 (S.E.); from day 23 to 40 the value was 12.1 $\pm$ 0.27 μg (Kincl et al. 1968).

Steroid-lipid mixtures can be made by melting steroids with various lipids and forming the mixture into pellets of various sizes. Dissolutions of steroids from such pellets are, in distilled water or saline again stable after an initial precipitous drop. In general it can be stated that for steroid-lipid pellets the dissolution of a steroid will be influenced by the composition of the mixture, the nature of the lipid phase used, and the surface area of the pellet.

Monitoring excreted radioactivity in urine and faeces of hamsters, rats and rabbits and determination of amounts circulating in peripheral plasma was used to measure the amounts of material released in vivo by sustained release preparations. Measurement of total excreted radioactivity provides only an approximate indication of absorptions since the parent steroid has been in most instances metabolized, and the metabolites may be also excreted by the skin, expired in breath and stored in various tissues. Excretion pathways may be different from species to species and from compound to compound.

The release of steroids from PDS implants and cholesterol pellets appears to follow roughly the same pattern. Figure 1 shows absorption of ^{3}H-19-norprogesterone from cholesterol pellets (20:80 mixture) and PDS implants monitored as total radioactivity in urine and faeces (Kincl et al. 1970a).

FIGURE 1

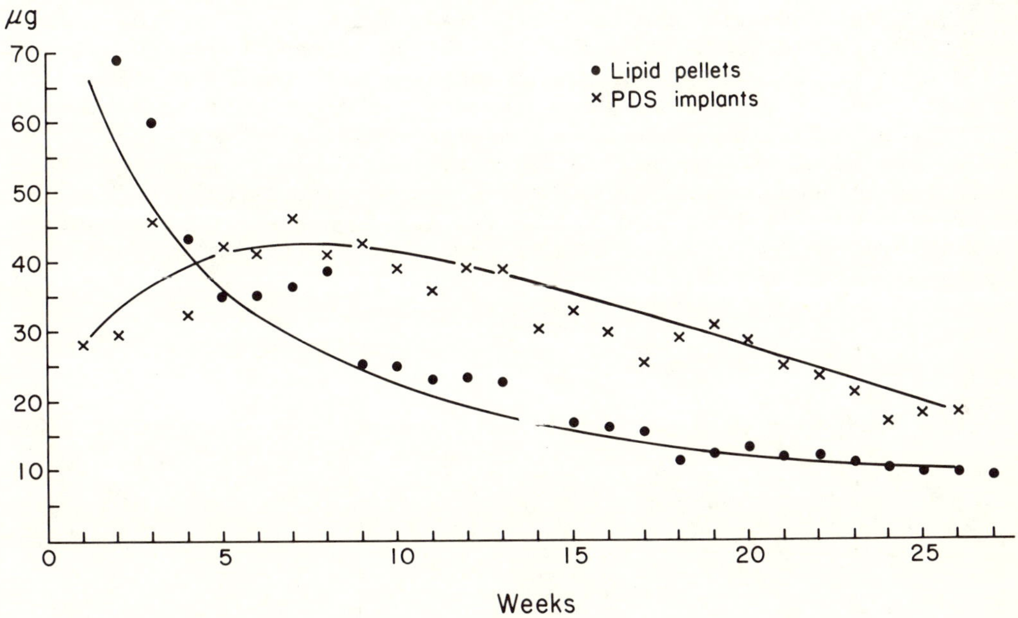

The initial value of excreted radioactive material was about 70 μg/day/pellet with lipid pellets. This declined to a mean of 9 μg/day at the end of 27 weeks. In animals having PDS implants excretion of radioactivity rose from an initial value of 32 μg/day/implant to 46 μg/day at the end of 3 weeks.

A lack of constancy was also seen in a six months study using megestrol acetate PDS implants in rats (Benagiano _et al_. 1970). In this study the general slope of the regression line was calculated to be -0.266.

The decline in release by either implant may be due to encapsulation noted in varying degrees with both steroid-lipid pellets and PDS capsules. In addition, PDS capsules may have another disadvantage. It was noted that capsules removed after a long-term use in humans had lower _in vitro_ release rates than originally (Benagiano, 1970). It could be that lipid-like material from the surrounding tissue was adsorbed by the polymer thereby altering the characteristics of the release. For the purpose of relatively short biological tests this lack of constancy is probably not important. The release of steroid during 2-3 weeks after implantation is often equivalent to _in vitro_ release (Benagiano _et al_. 1970).

Comparison of megestrol acetate plasma levels in rats after oral administration and released from PDS implants (21 μg/24h _in vitro_) is shown in Figure 2. The period of observation was 1 to 14 days after insertion for PDS implants and 1/2 to 24 h after oral administration (Kincl _et al_. 1970b). The doses used were equivalent based upon biological response. The data demonstrated that after an oral dose, which must be given in much higher amounts to produce a physiological effect than if a sustained release form is used, the plasma levels are significantly elevated. This in turn results in increased accumulation in various tissues (_vide infra_) which may impose stresses on liver and kidney functions, expose other organs to high levels of drug's metabolites and thus contribute to undesirable side effects.

FIGURE 2

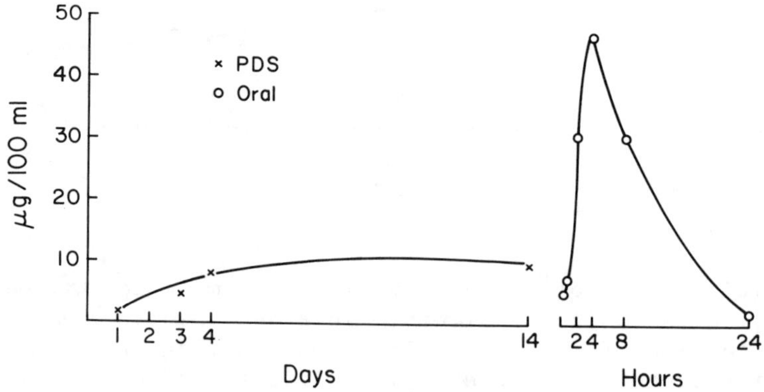

Permeation of steroids through polydimethylsiloxane membranes *in vitro* was shown to be proportionate to surface area and inversely proportionate to membrane thickness. When implanted in animals the conditions become obviously much more complex.

Membrane thickness appears to be less important. Plasma levels of megestrol acetate were higher in animals having implants made from a thicker wall tubing (larger diameter) than in animals with thinner wall tubes (Kincl and Rudel, 1970).

It appears from these studies that modalities controlling absorption *in vivo* will be the area, solubility of the compound in body tissues and transportation from the site of implantation, and possibly metabolic inactivation if the ability of the body to inactivate released substances is exceeded by the released amounts.

The uptake of radioactivity in key metabolic organs (liver, kidneys), adrenals, gonads and the brain in animals having megestrol acetate PDS implants were markedly lower than in animals receiving the same steroid by gavage when dosages by these two routes of administration were selected on the basis of biological potencies. Peak concentration after oral administration in the liver and kidneys was 40 times (liver) and 60 times (kidneys) higher than in animals with PDS implants. At the end of 24 h there was still greater uptake in the adrenals, the testes and the brain after oral administration as compared to a sustained release form (Kincl *et al*. 1970b).

The concept that a sustained release preparation may provide biological effect at lower dosages than a conventional once-a-day administration was tested in several biological tests using various

TABLE 1

RELATIVE BIOLOGIC POTENCY OF VARIOUS STEROIDS ADMINISTERED IN PDS IMPLANTS

Steroid used	Relative potency (subcutaneous injection = 100)
Testosterone	1500-3000
Progesterone	1250-2500
19-Norprogesterone[a]	2500
Norgestrel[b]	3200-4100
Megestrol acetate[c]	700-1340

[a] 19-nor-4-pregnene-3,20-dione

[b] dl-13β-ethyl-17α-ethynyl-17β-hydroxy-4-gonen-3-one

[c] 6-methyl-17α-acetoxy-4-pregnene-3,20-dione

Data from Chang & Kincl (1968;1970) and Zbuzkova and Kincl (1970)

steroids. It was shown that for several biologic effects smaller quantities are needed if the steroids are released continuously than when the same compound is given once-a-day by subcutaneous injection or orally. Increase in effectiveness may range from a 10-fold to a 40-fold increase (Table 1)

To summarize, the important feature of a sustained release form is the demonstration of increased biological effectiveness, and of therapeutic index. The observation that accumulation of a drug and/or of its metabolites was markedly decreased in many tissues when a sustained release form was used in contrast to an oral administration shows that such implants may act as an "artificial gland", the rate of release being nearer to the biological requirement. Possibly many drugs delivered in this manner may exhibit fewer "side effects".

REFERENCES

Benagiano G.: (1970) Personal communication.

Benagiano G., Ermini M., Chang C.C., Sundaram K., & Kincl F.A.: Acta endocr. (Kbh) 63 (1970) 29.

Chang C.C. and Kincl F.A.: Steroids 12 (1968) 689.

Chang C.C. and Kincl F.A.: Fertil. and Steril. 21 (1970) 134.

Dziuk P.J. and Cook B.: Endocrinology 78 (1966) 208.

Folch Pi A., Oriol A., Herra Lasso L., Maqueo M., Dorfman R.I. & Kincl F.A.: Acta endocr. (Kbh) 48 (1965) 602.

Kincl F.A., Folch Pi A. and Oriol A.: J. Reprod. Fertil. 12 (1966) 225.

Kincl F.A., Benagiano G., & Angee I.: Steroids 11 (1968) 673.

Kincl F.A., Sundaram K., Chang C.C. & Rudel H.W.: Acta endocr. (Kbh) (1970a) In Press.

Kincl F.A., Angee I., Chang C.C. & Rudel H.W.: Acta endocr. (Kbh) (1970b) In Press.

Kincl F.A. and Rudel H.W. (1970) To be published.

Lipschutz A., and Iglesias R.: Nature (Lond.) 190 (1961) 174.

Shimkin M.B. and White J.: Endocrinology 29 (1941) 1020.

Zbuzkova V. and Kincl F.A.: Endocr. exp. (1970) In Press.

EXPERIENCE WITH NORGESTREL IN CONTINUOUS MICRODOSAGE

GEORGE L. FOSS

MALE SUBFERTILITY AND ENDOCRINE

CLINICS, BRISTOL

Preliminary reports have been made on the use of dl-Norgestrel 50 micrograms in continuous daily doses as a contraceptive. Foss (1968), Foss et al (1969).

From February 1966 until October 1969, 242 women have been included in this trial and a total of 4278 cycles have been recorded. Except for a few newly married, most of these women had had 1 to 7 children. Their age distribution was from 17 to 48 years and some of them have completed 4 years on this regime. There was inevitably a considerable dropout. (Table 1)

Pregnancies

Of these 114 women, 13 became pregnant. Nine of these admitted missing tablets, either they forgot to take them on holiday or in some cases there was a death in the family.

Table 1. Dropouts.

Reason	Patients	
Mestrual irregularity, polymenorrhoea menorrhagia, oligomenorrhoea	33	13.6%
Wanted another child	30	12.4%
Moved from Bristol	15	6.2%
Pregnant	13	5.4%
Frightened by propaganda	6	2.5%
Various	17	7.0%
Total	114	

Two of these cases aborted. One patient had diarrhoea and vomiting when presumably this small dose of Norgestrel was not absorbed. Three of these women who became pregnant flatly denied having missed any tablets.

1. age 27, after 11 months with reasonable cycle control.
2. age 22, after cycles of 26,12,13 and 10 days, she left Bristol and it was not possible to obtain the true facts.
3. age 22, a very reliable patient who had taken microdosage for 36 months continuously.

One must accept three pill failures which gives a pregnancy rate of 0.84/100 w.y. and the overall pill and patient failure gives a pregnancy rate of 3.6/100 w.y.

Cycle Control

The largest number of patient gave up this regime or were put on to some other method because of menstrual irregularity, repeated or very frequent short cycles, with bleeding which was in no way different from their normal menstrual pattern. More rarely the interval was so prolonged that they became worried. This method did not suit 13.6% or 1 out of 7 patients for this reason. The so called normal woman has been investigated by Chiazze et al (1968), who recorded 30,655 cycles from 2316 American and Canadian volunteers and they found that 83% of cycles were between 23 to 33 days, whereas in the Norgestrel series for the same range, the figure was 68% and figures 1 and 2 show that the 22% of shorter cycles occur around the 14th and 15th days.

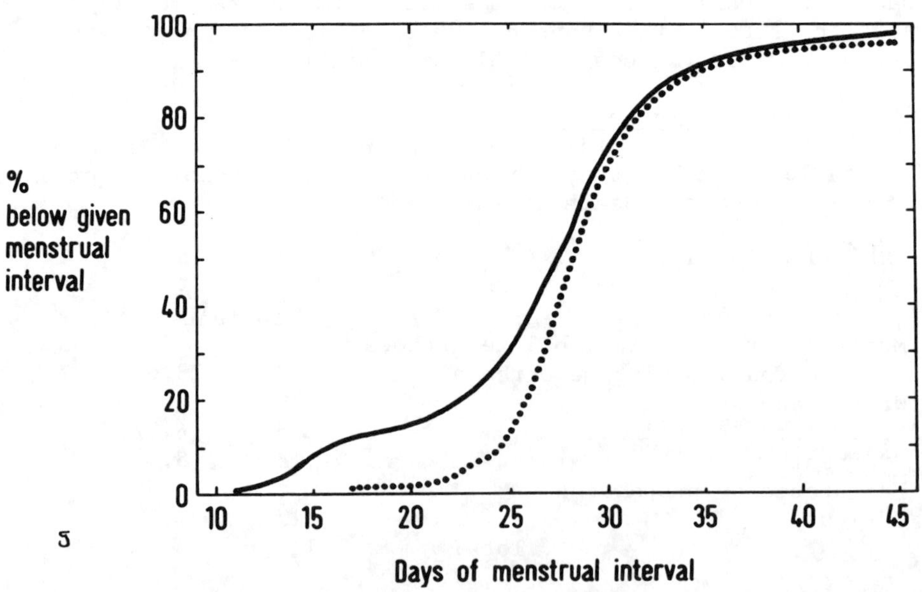

Fig.1 ---- Menstrual cycle variation on 50 ug Norgestrel, 4,278 c.
.... Chiazze et al., 1968. Normal women 30,655 cycles

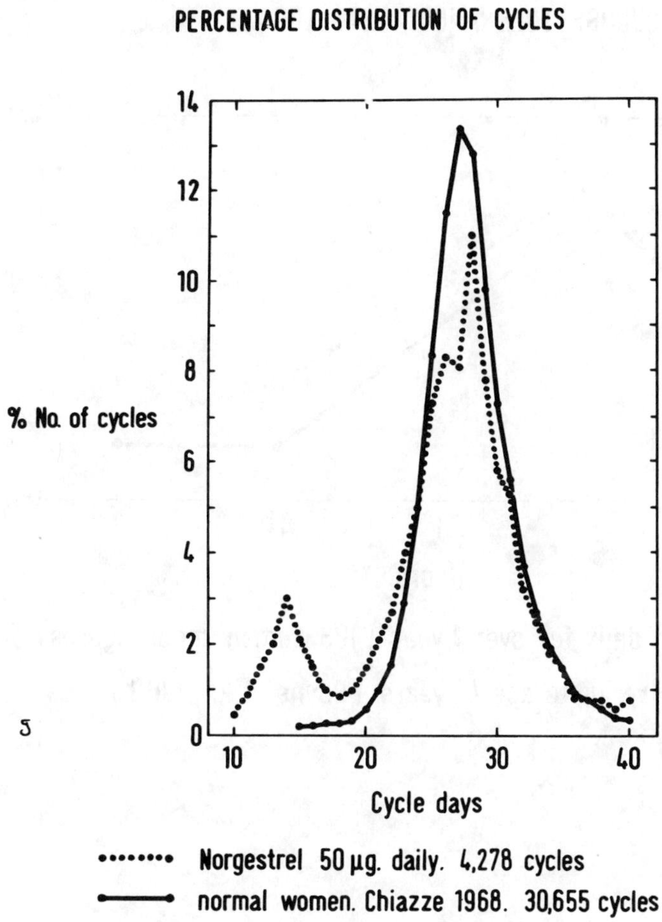

Fig.2

The percentage of longer cycles is the same as in normal women. It is only when an individual has recurrent short cycles that the method is rejected. The average length of all the 4278 cycles was 26.636 days (SD 8.323) compared with Chiazze's 29.1 (SD 7.46). As reported previously side effects are virtually non-existant - 5% of women at some time admitted having breast tenderness. There was no definite trend in alteration of weight.

Glucose Tolerance

After being on continuous Norgestrel for over 1 year and up to 3 years, oral glucose tolerance tests were carried out on 103 cases using 50G glucose (Foss et al 1970). The mean results were compared with those of a group of 64 normal women of the same age group, investigated by Boyns et al (1969). There was slightly greater tolerance in the Norgestrel treated women. (Fig 3)

ORAL GLUCOSE TOLERANCE TESTS (50g.)

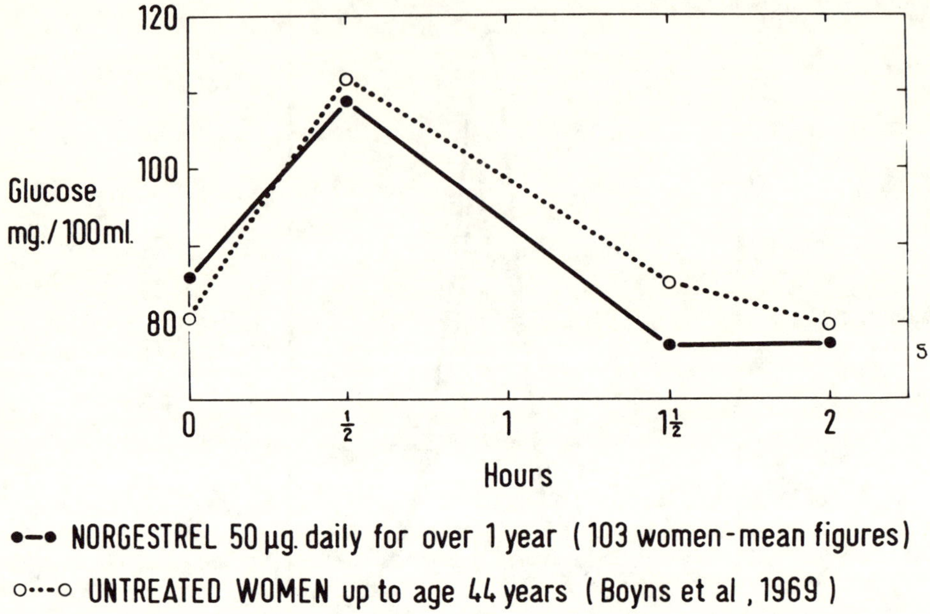

Fig.3

When the means were plotted according to the day of the cycle when the test was conducted, it was seen that tolerance became less as the cycle progressed and this was a similar finding in normal untreated women by Jarrett and Graver (1968) and is interpreted as further evidence that Norgestrel produces no impairment of glucose tolerance (Fig. 4)

Liver function. As a screening test in these 103 patients, blood was also taken for estimation of serum glutamic-pyruvate transaminase (SGPT) and all the figures were within normal limits - mean 30 (SD 10) controls 35 (SD 12.5)

Thyroid function. Blood was investigated in a series of 50 women who had been on this method of contraception for over 6 months by Goolden et al (1970). The parameters used were Protein bound Iodine, serum thyroxine-iodine, (125I) -T3 sephadex uptake and T.B.G. binding capacity. Their findings showed that thyroid function was unaffected by Norgestrel.

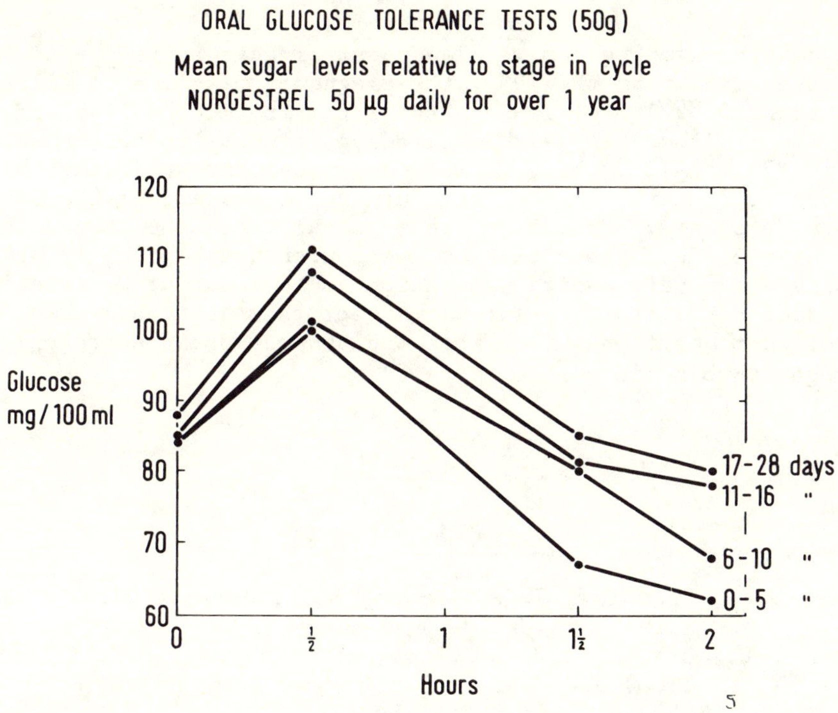

Figure 4.

Mode of Action

These minute doses have been shown to produce atypical endometrial appearances at all stages of the cycle (Foss 1968). It is probable that ovulation is not inhibited but there is some evidence that 50µg Norgestrel interferes with luteal function and lower than normal values of pregnanediol were found. (Fotherby et al 1968). More recently Wright et al (1970) showed that Norgestrel does not interfere with corpus luteum development. However, values for pregnanediol excretion between the 19th and 24th days of the cycle were lower than those found during a normal menstrual cycle of a woman not taking Norgestrel. It was suggested that possibly Norgestrel may interfere with the metabolism of progesterone or decrease its synthesis or secretion. Changes in cervical mucus are prominent and this is difficult to collect, but actual movement of spermatozoa does not appear to be inhibited.

Conclusions

In continuous 50µg dosage dl-Norgestrel gave better protection against pregnancy than the IUCD, the pregnancy rate due to pill failure being 0.84/100 w.y. and a total pill/patient failure of 3.6/100 w.y. Although mid-cycle bleeding, similar to menstruation, occurs in 22% of all cycles, it was only recurrent and enough of a nuisance to make this method of contraception unacceptable to 1 out of 7 patients. Because of this and also because it does not give as near perfect protection as the conventional pill, it has not received adequate medical attention. The 6 out of 7 patients whom it does not disturb prefer and accept this apparently harmless method of contraception rather than the conventional oestrogen/progestogen combination.

References

1. Boyns, D.R., Crossley, J.N., Abrams, M.E., Jarrett, R.J. & Keen, H. (1969) Brit.Med.J. 1, 595.
2. Chiazze, L., Brayer, F.T., Macisco, J.J., Parker, M.P. & Duffy, B.J. (1968) J.A.M.A., 203, 377.
3. Foss G. L. (1968) J. Reprod.Fert.Suppl.5. 145.
4. Foss, G.L., Svendsen, K.K., Fotherby, K. & Richards, D.J. (1968) Brit.Med.J. 4, 489.
5. Foss G.L., Holton, J.B. & Lewis, F.J.W. (1970) in the Press.
6. Fotherby, K., Svendsen, E.K. & Foss, G.L. (1968) J. Reprod. Fert. Suppl.5, 155
7. Goolden, A.W.G., Bateman, D.M., Pleehachinda, R., & Sanderson, C. (1970) Lancet, 1, 624.
8. Jarrett, R.J. & Graver, H.J. (1968) Brit. Med.J. 2, 258.
9. Wright, S.W., Fotherby, K. & Fairweather, F. (1970) J. Obstet. & Gynaec. Brit. Comwlth. 77, 65.

INFLUENCE OF LOW DOSES OF GESTAGENS ON SOME PITUITARY-OVARIAN FUNCTIONS

Ulf Larsson-Cohn

Department of Obstetrics and Gynaecology

University Hospital, Uppsala, Sweden

The idea of using low doses of gestagens as oral contraceptives (OC) was first explored by Rudel et al. (19). In recent years several reports have appeared describing the clinical results during this kind of treatment, for a summary see (9). Some authors (3, 18) seem to consider the major contraceptive effect of these drugs to be on the cervical mucus, making it hostile to spermatozoa while other investigators (6, 15, 22) believe that this can not be the only explanation.

We have, at the University Hospital of Uppsala, Sweden, performed a series of studies aimed to investigate the effects of low doses of gestagens on some pituitary-ovarian functions (10-14). In each subject the investigation period consisted of one control cycle followed by two or three months of treatment with a low daily dose of a gestagen. The following drugs were used: Norethindrone (NET) 0.5 mg, 0.3 mg and 0.1 mg, Chlormadinone acetate (CMA) 0.5 mg, d-Norgestrel 0.03 mg and the retroprogesterone Ro 6-3129 4 mg. The daily excretion of oestrogens was measured in all studies. Using the method of Johansson (8) the plasma concentration of progesterone was determined daily during the luteal phase of the cycles and in general every third day during the rest of the period of investigation. Daily determinations of the urinary excretion of luteinizing hormone (LH) were performed in three of the studies (10, 11, 13) while the excretion of pregnanediol was measured in two of them (10, 12). Figs. 1-6 show some results from three of these investigations (10, 11, 12). The results may be summarized as follows:

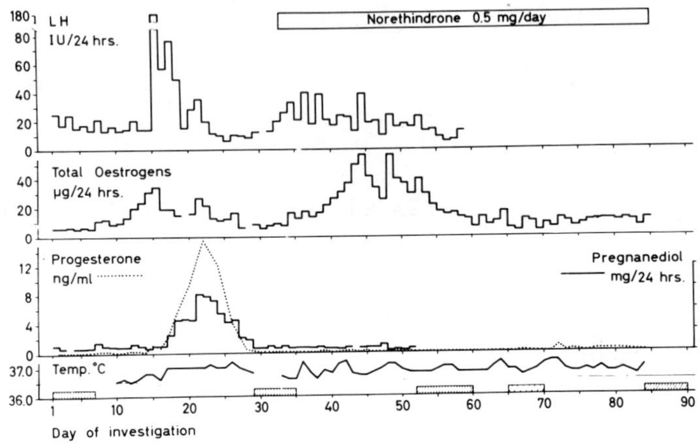

Figure 1. Effects of 0.5 mg daily of Norethindrone of the plasma levels of progesterone and on the urinary excretion of LH, pregnanediol and oestrogens (10).

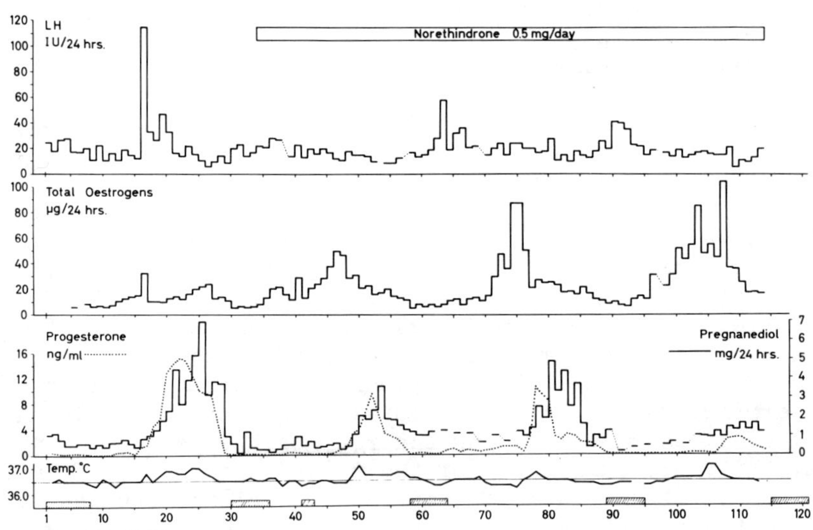

Figure 2. See figure 1.

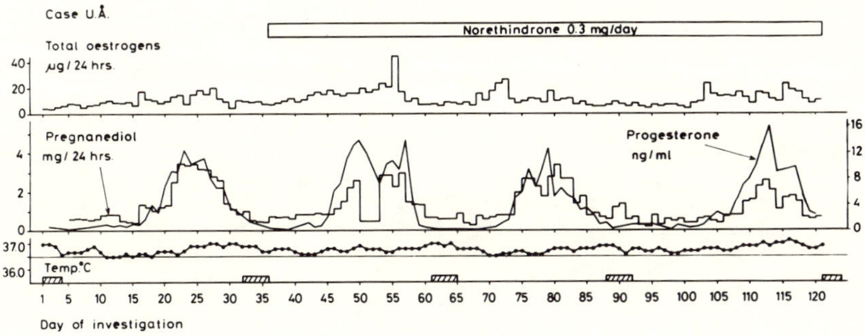

Figure 3. Effects of 0.3 mg of Norethindrone on the plasma levels of progesterone and on the urinary excretion of pregnanediol and oestrogens (12).

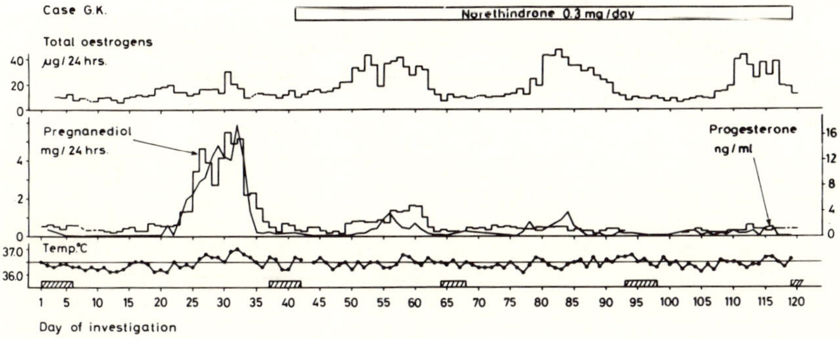

Figure 4. See figure 3.

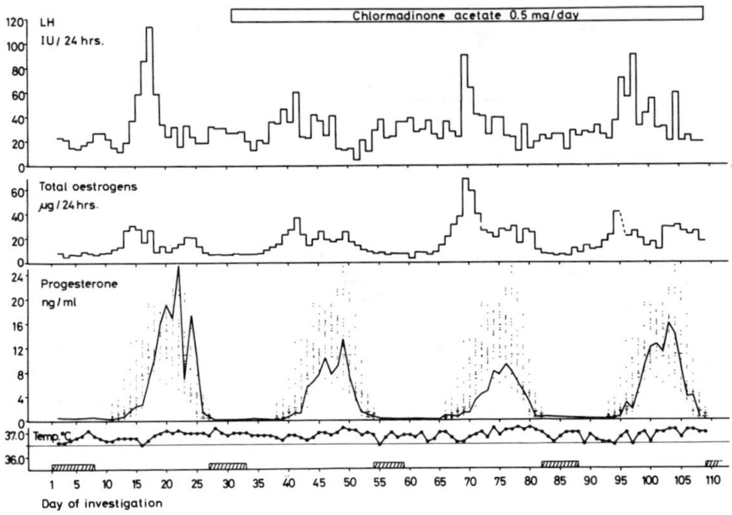

Figure 5. Effects of 0.5 mg of Chlormadinone acetate on the plasma levels of progesterone and on the urinary excretion of LH and oestrogens (11).

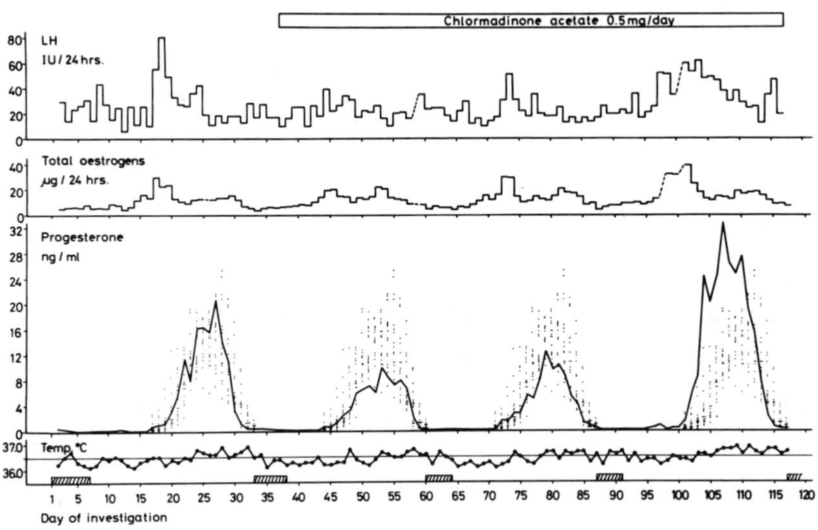

Figure 6. See figure 5.

Urinary excretion of LH.

When the drug was Ro 6-3129 the excretion pattern was normal except that there seemed to be a tendency to early appearence of the midcycle peak (13). It was speculated that this drug possibly acted as an ovulatory stimulater, as does the closely related retrosteroid Ro 4-8347, and that when such a drug is administered to a normally menstruating woman it may shorten the duration of the follicular phase of the cycle. The LH excretion was also determined in the studies where the drug was 0.5 mg of NET or CMA. It was found that in most cases the midcycle peak disappeared while the basal levels were normal. The influence of NET was stronger than that of CMA. It was concluded that one effect of these steroids is to distort the regulation of the LH secretion rather than to suppress it. These results partly agree with earlier investigations where the urinary and plasma levels of LH were determined during treatment with CMA 0.5 mg (2, 7, 16).

Urinary excretion of oestrogens.

All women studied had a cyclical oestrogen excretion pattern, also those with consistantly low progesterone and pregnanediol levels. The total amount of excreted oestrogens was increased in many of the subjects, this was most apparent in women treated with 0.5 mg of NET or CMA. There was in serveral cases a tendency to an indirect relationship between the oestrogen and the progesterone levels.

The raised excretion of oestrogens was unexpected. Previous studies of the levels of urinary oestrogens during treatment with NET (10) or CMA (11) have shown either decreased or unchanged levels. Only up to 0.6 per cent of orally administered NET is excreted as oestrogens (10), and there is no indication that CMA is metabolized to substances with oestrogenic properties (11). It is therefore unlikely that the elevated values were significantly affected by oestrogenic metabolites of CMA or NET being assayed.

The mechanism repsonsible for the elevated oestrogen excretion is unknown. However, Stevens & Vorys (20) reported that 0.25 and 0.5 mg of Ethynodiol acetate stimulated the FSH excretion and that the lower dose did so to a greater extent that the larger one. It can be speculated that also low doses of CMA and NET may raise the FSH secretion. This could possibly cause multiple follicle development with an increased oestrogen production as a consequence. Another explanation could be that the medication induced some alterations of the activity of the enzymes responsible for the ovarian steroid production.

Plasma concentration of progesterone and urinary excretion of pregnanediol.

With the exception of the cycles during which conception occured, the progesterone concentration varied within normal ranges in the subjects treated with NET 0.1 mg and Ro 6-3129 4 mg. As illustrated in figs. 1-6, the luteal activity was normal, sligthly depressed or highly depressed in women taking NET 0.5 mg, NET 0.3 mg or CMA 0.5 mg. There is no generally accepted standard for what should be considered to be a normal luteal progesterone complex. In a study of 20 apparantly ovulatory menstrual cycles, each cycle showed at least one progesterone value higher than 12 ng/ml (8). It therefore seems that a luteal complex where the progesterone concentration never exceeds the arbitrarily selected level of 10 ng/ml could be considered as abnormal. Table 1 shows the number of cycles during treatment with NET 0.5 mg, NET 0.3 mg and CMA 0.5 mg in which all progesterone values were below 10 ng/ml. It also gives the number of cycles where the plasma levels either were consistantly on a follicular level or only showed a very moderate increase of short duration.

Drug	No of treatment cycles	Cycles with plasma progesterone levels never exceeding 10 ng/ml		Cycles with strongly depressed plasma progesterone levels	
		No	%	No	%
NET 0.5 mg	9	8	89	6	67
NET 0.3 mg	20	9	45	6	30
CMA 0.5 mg	18	5	28	0	0

Table 1 Number of cycles with plasma progesterone levels never exceeding 10 ng/ml and number of cycles with strongly depressed progesterone levels.

During treatment with d-Norgestrel 0.03 mg, the results were fairly similar to those obtained when the drug was CMA 0.5 mg.

In most cases there was a good correlation between the plasma progesterone levels and the urinary excretion of pregnanediol. Reduced pregnanediol levels has earlier been reported in women treated with 0.5 mg of CMA (2, 5, 15, 22), 0.1 mg of NET (2) and 0.05 mg of Norgestrel (4). In a study by Østergaard and Starup (17), later confirmed by Wright et al. (21), women on gestagens were found to have macroscopically and microscopically normal corpus luteum concomitant with a low excretion of pregnanediol. It was suggested that a defective function of the corpus luteum was the explanation to this discrepancy.

The fact that in the present series of studies almost all periods of elevated progesterone and pregnanediol levels were maintained for a period corresponding to the duration of the luteal phase of a normal menstrual cycle supported the theory that ovulation occured. It therefore seemed plausible that the low levels reflected an altered function of corpus luteum. This could possibly be due to a decreased synthethis of progesterone, as reported to occur *in vitro* in the presence of CMA (1).

It is doubtful if normal conception and/or nidation may occur in menstrual cycles with very low luteal progesterone concentration. The clinical significance of moderately decreased luteal activity is unknown, although it does not seem unlikely that even small disturbancies of the normal hormone balance may reduce the fertility. It can be concluded that the contraceptive mechanism of low doses of gestagens is incompletely understood, but that several factors, including defect corpus luteum function and alterations of endometrium and cervical mucus probably are of importance.

References

1. Aakvaag, A.: Formation of steroid hormones in the porcine ovary in vitro. J. Endocr. 43:25 (1969).
2. Diczfalusy, E., Goebelsmann, U., Johannisson, E., Tillinger, K.-G., and Wide, L.: Pituitary and ovarian function in women on continuous low dose progestogens; effect of Chlormadinone acetate and Norethisterone. Acta Endocr. (Kbh.) 62:679 (1969).
3. Elstein, M.: Urinary hormone excretion patterns in Chlormadinone acetate treated patients and the associated cervical mucus changes. In *Chlormadinone acetate, a new departure in oral contraception.* Editors: G.A. Christie and M. Moore-Robinson. Excerpta med. (Amst.) 1969, p. 46.
4. Foss, G.L., Svendsen, E.K., Fotherby, K., and Richards, D.J.: Contraceptive action of continuous low doses of norgestrel. Brit. med. J. 4:489 (1968).
5. Geller, G.: In *Chlormadinone acetate, a new departure in oral contraception.* Editors: G.A. Christie and M. Moore-Robinson. Excerpta med. (Amst.) 1969, p. 108.
6. Gibor, Y., Cohen, M.R., and Scommegna, A.: Effect of continuous administration of small doses of Chlormadinone acetate on the cervical mucus and postcoital test. Fertil. Steril. 20:572, (1969).

7. Jaffe, R.B., and Midgley, A.R. Jr.: Effects of synthetic steroids on serum gonadotropins. In Progress in Conception Control 1968. Editor: D.L. Moyer, J.B. Lippincott, Philadelphia 1968, p. 13.

8. Johansson, E.D.B.: Progesterone levels in peripheral plasma during the luteal phase of the normal human menstrual cycle measured by a rapid competitive protein binding technique. Acta endocr. (Kbh.) 61:592 (1969).

9. Larsson-Cohn, U.: Contraceptive treatment with low doses of gestagens. Acta endocr. (Kbh.) Suppl. 144 (1970).

10. Larsson-Cohn, U., Johansson, E.D.B., Wide, L., and Gemzell, C.: Effects of continuous administration of 0.5 mg of Norethindrone on the plasma levels of progesterone and on the urinary excretion of luteinizing hormone, pregnanediol and total oestrogens. Acta endocr. (Kbh.) 63:216 (1970).

11. Larsson-Cohn, U., Johansson, E.D.B., Wide, L., and Gemzell, C.: Effectsof continuous administration of 0.5 mg of Chlormadinone acetate on the plasma levels of progesterone and on the urinary excretion of luteinizing hormone and total oestrogens. Acta endocr. (Kbh.) 63:705 (1970).

12. Larsson-Cohn, U., Johansson, E.D.B., and Gemzell, C.: Effects of continuous daily administration of 0.3 mg of Norethindrone on the plasma levels of progesterone and on the urinary excretion of pregnanediol and total oestrogens. Acta endocr. (Kbh.) 64:38 (1970).

13. Larsson-Cohn, U., Johansson, E.D.B., Wide, L., and Gemzell, C.: Effects of daily administration of a retrosteroid on the plasma levels of progesterone and on the urinary excretion of luteinizing hormone and total oestrogens. J. Obst. Gynec. Brit. Cwlth. in press.

14. Larsson-Cohn, U., Johansson, E.D.B., and Gemzell, C.: Unpublished.

15. Martinez-Manautou, C., Giner-Velazquez, J., Aznar-Ramos, R., Lozano-Balderas, M., and Rudel, W.H.: Continuous administration of 0.5 mg of Chlormadinone acetate as a method of regulating fertility without inhibiting ovulation. In Proceedings of the eight international conference of the International Planned Parenthood Federation Santiago Chile 9-15 april 1967. Editors: R.K.B. Hankinson, R.L. Kleinman, P. Eckstein, and H. Romero. IPPF, London, 1967, p. 241.

16. Orr, A.H., and Elstein, M.: Luteinizing hormone levels in plasma and urine in women during normal menstrual cycles and in women taking combined contraceptives or Chlormadinone acetate. J. Endocr. 43:617 (1969).

17. Østergaard, E., and Starup, J.: Occurrence and function of corpora lutea during different forms of oral contraception. Acta Endocr. (Kbh.) 57:386 (1968).

18. Roland, M.: Norgestrel-induced cervical barrier to sperm migration. J. Reprod. Fert. Suppl. 5:173 (1968).

19. Rudel, H.W., Martinez-Manautou, J., and Maqueo-Topete, M.: The role of progestogens in the hormonal control of fertility. Fertil. Steril. 16:158 (1965).

20. Stevens, V.C., and Vorys, N.: The regulation of pituitary function by sex steroids. Obst. Gyn. Survey 22:781 (1967).

21. Wright, S.W., Fotherby, K., and Fairweather, F.: Effect of daily small doses of norgestrel on ovarian function. J. Obstet. Gynaec. Brit. Cwlth. 77:65 (1970).

22. Zanartu, J., Rodriguez-Moore, G., Pupkin, M., Salas, O., and Guerrero, R.: Antifertility effect of continuous low-dosage oral progestogen therapy. Brit. med. J. 2:263 (1968).

ON THE WAY TO A MORE PHYSIOLOGICAL HORMONAL CONTRACEPTION

R. Štěrba

Research Institute of Pharmacy and Biochemistry

Prague, Czechoslovakia

Several authors (Jackson 1968, Heinen 1968) recommended to elevate the oestrogen doses in hormonal contraceptive preparations to 100 μg daily at least to enhance the contraceptive reliability; on the other hand, a number of authors (e.g., Rudel and Kinzl 1968, Martínez-Manautou 1968) recommended to completely omit the oestrogens and to administer progestagens alone at continual low dosages or (Tyler 1966) in single sustained-action injections.

Arguments against the continual administration of progestagens alone at low dosage levels by any of these routes include the irregularity of cycles, a considerable incidence of spotting, and high pregnancy rates, reported by certain authors (Butler 1969, Knaus 1969, Müller and associates 1969).

Arguments against the elevation of oestrogen dosages are based on the changes in the blood protein spectrum (Laurell 1969), on the elevation of blood glucose levels (Svoboda and Štěrba 1969), and on the rise in blood coagulability (Halberstadt 1968, Wessey and Doll 1968). According to Schaffoy and Scheinberg (1966), even cerebral vascular infarction occurred; hypertension accompanied by severe headaches is frequent (Leragh 1967, Woods 1967). An augmented retention of extravascular fluid after oestrogens may even lead to papillary congestion signalling central oedema (Arbenz and Wormser 1965).

For these reasons, new oestrogen-progestagen combinations have been recently strongly preferred, containing low doses of so-called pure progestagens of the acetoxyprogesterone type, whose activity is enhanced by halogenation.

In the years 1967 - 1969, a biphasic (sequential) contraceptive preparation, Biogest, was being developed at the Research Institute of Pharmacy and Biochemistry in Prague. In the initial formulation, for the first eleven-day phase oestrogen alone, and for the second ten-day phase, a combination of oestrogen and a halogenated progestagen was provided. The latter was at first represented by Chlorsuperlutin (6-chloro-6-dehydro-16-methylene-17alpha-acetoxyprogesterone); later on, also the corresponding 6-bromo analogue Bromsuperlutin was tested.

Since Řežábek (1969) pharmacologically documented a high antioestrogen activity particularly of Chlorsuperlutin, and since we ourselves brought clinical evidence about the dissociation of the progestative and antioestrogen activities in halogenated progestagens of the acetoxyprogesterone type, we changed the formulation and prepared a new combined and simultaneously sequential contraceptive. For the first phase of the cycle, each dose contained 0.08 mg mestranol with 0.25 mg of either Chlor- or Bromsuperlutin added, and for the second phase, each dose contained 0.08 mestranol again, combined with 1 mg of either Chlor- or Bromsuperlutin. The dose of 0.08 mg mestranol may be reduced only if the progestagen added to it belongs to the norsteroid group, since these latter compounds are converted to oestrogens in the organism (Kaiser 1960, Breuer 1960, McGinty and Djerassi 1958, Brown and associates 1960), whereby the total oestrogenic activity of the preparation is enhanced in addition. On the other hand, the dose of 0.08 mg mestranol is likely to be set as the upper admissible limit by the World Health Organization to eliminate potential hazards.

Our new formulation of the contraceptive preparation (Tab.1) was expected to produce the following effects:
1. Regular cycles, characteristic of a biphasic preparation, free from spotting and breakthrough bleeding.
2. A Sufficient hypophysary blockade safeguarded by an adequate dosage of oestrogen.
3. A distinct effect (recently contested again by Müller and associates 1968, Knaus 1968) exerted by low doses of progestagen upon the permeability of the cervical mucus for spermia within the very first phase of the cycle.
4. An influence on the FSH : LH ratio and on the endometrial nidation readiness.

Thus, in theory, the very last side effects of the sequential preparation should be eliminated, such as slight nausea and enhanced secretion from the cervical glands, or fluid retention during the first oestrogen phase of the cycle.

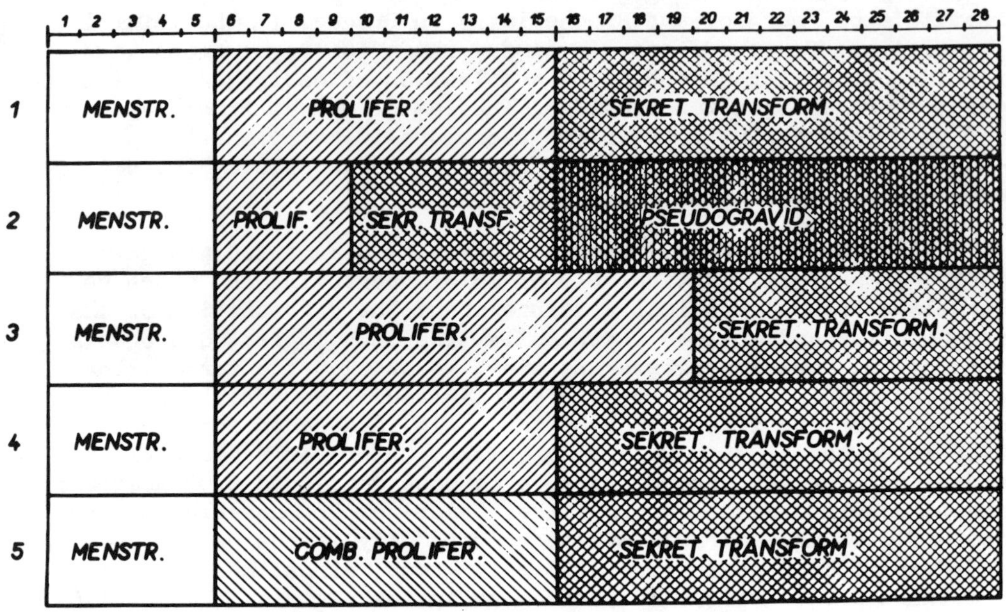

Table 1. Development stages of Biogest

1. Normal menstrual cycle exhibiting within a 28-day period the menstruation, proliferation, and secretion phases.
2. Cycle during usage of combined preparations, with both the proliferation and secretion phases shortened. From the 15th day onwards the so-called pseudogravidity phase sets on, during which spotting or even breakthrough bleeding may frequently occur. If the woman discontinues the medication when the breakthrough bleeding occurs, then the latter passes over to withdrawal bleeding, and consequently premature menstruation takes place.
3. Cycle during usage of biphasic preparation containing oestrogen (15 tablets) for the first phase, and oestrogen combined with progestagen (5 tablets) for the second phase.
4. Cycle during usage of biphasic preparation (11 + 10 tablets) with oestrogen phase shortened and combination phase extended against case 3.
5. Biological cycle during usage of preparation uniting the favourable properties of previous preparations. During the proliferation phase, a combination of mestranol with a low dose of progestagen, and during the second combined phase, a combination of mestranol with a higher dose of progestagen are administered respectively.

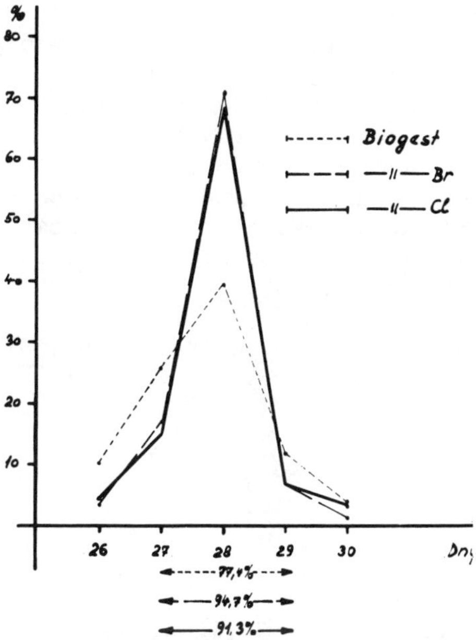

Graph 1

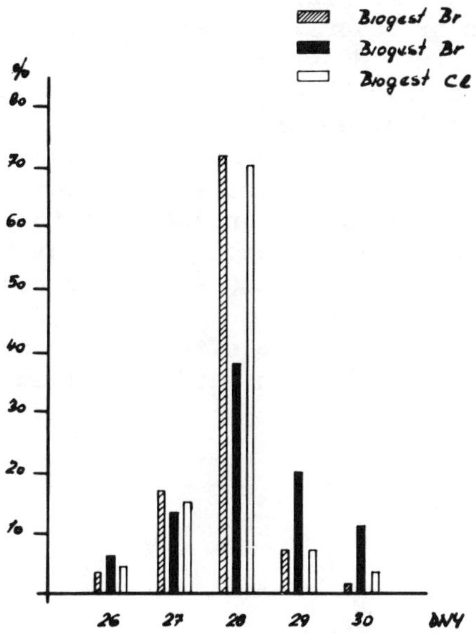

Graph 2

Distribution of cycle durations during testing of Biogest. The modification containing chloro progestagen is distinctly superior to that containing bromo progestagen, which in a higher total number of cycles exhibits markedly worse values.

Since even in a normal feminine cycle progesterone is never completely absent during the first phase, the modification described not only suffers no loss but on the contrary gains an improvement of its biological composition.

Clinical experiments were made with both modifications of Biogest, i.e., those containing either the chloro or the bromo derivative of the progestagen. From Graphs 1 and 2 it is evident that Chlorsuperlutin was superior to Bromsuperlutin.

When both of these modifications of combined contraceptives, containing halogenated progestagens, were being administered altogether to 284 women during a total number of 1236 cycles (i.e., for about a year), not a single unwanted pregnancy occurred. The pregnancy rate of Biogest, however, has to be established from the results of the present extended multi-centre research, being conducted at 20 clinical and institutional investigation centres and involving about 10,000 cycles.

A comparison with foreign preparations Stediril and Metrulen taken by several patients of ours during the preceding year, showed that after Biogest, in comparison with Stediril, the incidence of weak cyclic bleeding is higher and that of breakthrough bleeding is about seven times lower, the other indices being roughly equal. With regard to the low frequency and intensity of side effects and to the high regularity of cycles, Biogest may be ranged among the top preparations manufactured in the world. Likewise, reports from home clinical investigation centres emphasize the outstanding tolerance of Biogest and a perfect regularity of cycles.

No major side effects have ever been observed, although special attention has been paid to not only peripheral but also cerebral vessels.

References

1/ Arbenz, J.P., Wormser, P.: Schweiz.med.Wschr. 95,1654,1965
2/ Breuer, H. et al.: Acta endocr. (Kbh) 33,1960,433
3/ Brown, J.B., Blair, H.A.F.: Proc.roy.Soc.Med. 53,1960,433
4/ Butler, Ch., Hill, H.: Lancet 1,1116,1969
5/ McGinty, D.A., Djerassi, C.: Ann.N.Y.Acad.Sci. 71,1968,500
6/ Halberstadt, E.: Med.Wschr. 806-813,1968
7/ Heinen, G.: Die medizinische Welt 44,2420,1969
8/ Jackson, J.L. et al.: Fertility & Sterility 19,649,1968
9/ Kaiser, R.: Dtsch.med.Wschr. 85,1960,1457
10/ Knaus, Th.: Ref.: Zbl.f.Gynäk. 91,1051,1969
11/ Laurell, C.B. et al.: Clin.chim.Acta 35,294,1969
12/ Leragh, J.H. et al.: J.Amer.med.As. 201,918,1967
13/ Martinez-Manautou J.: Fertility & Sterility Exc.med.Found. 999,1968
14/ Müller, R. et al.: Ref. Zbl.f.Gynäk. 91,1050,1969
15/ Rudel, H.W., Kinzl, J.: IPPF Medical Buletin 2,5,1968
16/ Řežábek, K., Štěrba, R.: Čas.lék.čes., in press
17/ Schafoy, Scheinberg: Ref.: JAMA 1967
18/ Svoboda, Z., Štěrba, R.: Metabol.clin.and exper.New York, in press
19/ Tyler E.T.: Obstet. Gynec. 28,787,1966
20/ Wessey, M.P., Doll, R.: Brit.Med.J. 1968/II p.199
21/ Woods, J.W.: Lancet 1967/II 653

THE PILL : ON CHROMOSOMES

L. Gedda, G. Torrioli-Riggio, G. Brenci, L. Silvani
"G. Mendel" Institute for Medical Genetics
and Twin Research, Rome (Italy)

Chromosome studies in women who use the pill were suggested on the basis of statistical data concerning female fertility at various ages. A research study based on 5,000,000 births in Italy in the years 1962-1965 evidences the increase in natimortality with increased maternal age (cf. Fig. 1). The causes of natimortality are largely of a genetic nature, as revealed by a comparison between natimortality and neonatimortality. Further investigations show in fact that neonatimortality declines over the last century (as a result of improved environmental conditions), while natimortality remains virtually unchanged (as expected in the hypothesis of genetic conditioning).

The exponential increase of natimortality with increased maternal age indicates that the age of the gamete affects the expression of latent genetic defects. Thus aging of the gametes increases the probability that a parent will transmit

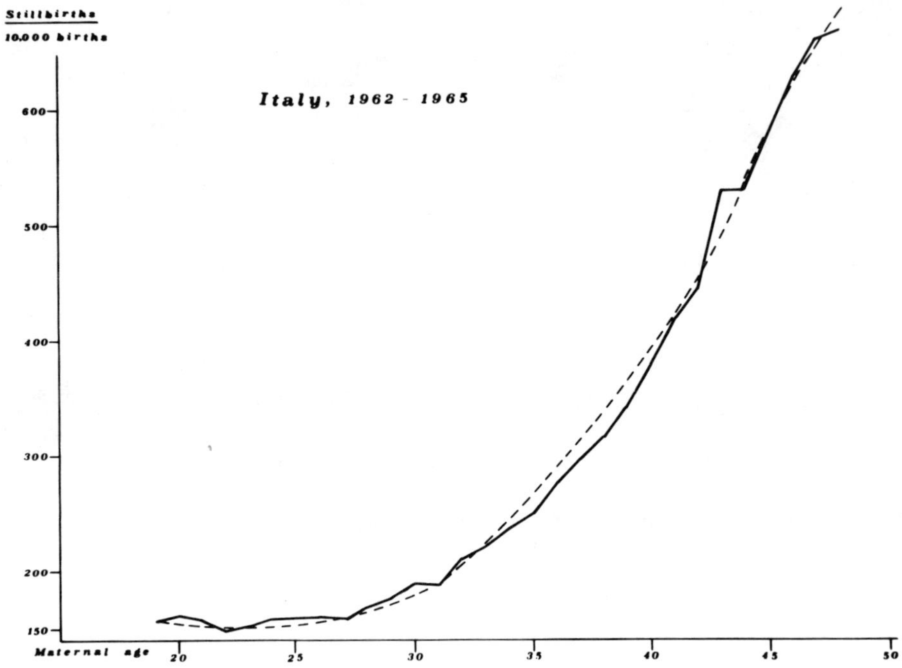

Fig. 1

more serious genetic defects. By blocking the maturation of the follicle, the pill induces aging of the ovum.

On the other hand, studies by Carr (1969) on stillbirths in women who take the pill suggest that aging of female gametes may be the result of damage to its chromosome complement. Carr found only 3 normal karyograms in 11 cases of stillbirths in women using oral contraceptives, while 5 showed polyploidy and 3 Turner's syndrome.

Since part of the genetic control is common to both mitosis and meiosis, we decided to verify the influence of the pill on the chromosomes of peripheral blood lymphocytes, applying standard cytogenetic techniques. Our study concerns 11 healthy women who used oral contraception (6 cases are from Stockholm, obtained with the precious cooperation of Prof. Ingelman-Sundberg to whom we are indebted; 5 cases are from Rome). The same technique was adopted in the study of the karyologic pattern in 11 untreated women, of corresponding age distribution, in the intermenstrual period. The results can be found in the Tables.

In Tab. I we find a comparison of quantitative analytical data as provided by the experimental and control samples, respectively.

The first finding is a high degree of polyploidy in the experimental sample but not in the control sample. As for diploid plates, the experimental sample shows an extension to the left of the distribution of the chromosomes (i.e. increased numbers of hypoploid plates, with 42 through 45 chromosomes), while

Tab. I

	EXPERIMENTAL SAMPLE									CONTROL SAMPLE								
		Polyploid	DIPLOID PLATES								Polyploid	DIPLOID PLATES						
N.	Cases	Plates ‰	n. of Plates	Plates by Chromos. numb. 42	43	44	45	46	47	Cases	Plates ‰	n. of Plates	Plates by Chromos. numb. 42	43	44	45	46	47
1	C.E.	16	29	-	-	1	4	22	2	B.P.	-	29	-	-	1	2	24	2
2	E.K.	6	15	1	1	-	2	11	-	B.L.	-	30	-	-	1	1	27	1
3	A.L.	3	22	-	1	1	1	16	3	M.I.	-	28	-	-	-	3	25	
4	F.H.	14	21	-	-	-	1	20	-	D.F.F.	-	21	-	-	1	1	18	1
5	G.H.	7	31	1-	-	-	5	23	2	N.R.	1	22	-	-	1	2	19	
6	K.S.	18	28	2	1	1	5	17	2	T.L.	-	26	-	1	-	1	23	1
7	N.L.	-	18	1	1	3	1	12	-	L.L.	-	32	-	-	2	1	28	1
8	R.A.	-	16	-	-	1	2	12	1	F.A.	1	23	-	-	-	1	21	1
9	L.H.	6	15	-	-	1	5	9	-	M.M.	-	25	-	-	1	1	22	1
10	S.S.	5	15	-	1	1	4	7	2	M.L.	-	30	-	-	2	3	24	1
11	M.A.	3	17	-	-	2	1	14	-	D.F.R.	-	30	-	-	1	3	26	
	Total	7.6 *	227	5	5	11	31	163	12	Total	0.2 *	296	- -	1	10	19	257	9

* mean value ‰

the distribution towards the right is the same for both samples.

Tab. II presents again the data of the previous table, concerning our experimental and control samples, rearranged in order to provide a comparison with standard data (taken from Bartalos and Baramki, 1967).

As a first attempt at qualitative analysis of our data, we can only assess, for the time being, a slight, nonspecific increase in the number of chromosomal alterations (atypically long or short arms, breaks, translocations, associations) reminding us of the observations made by Hsu and Somers (1961) on mammalian chromosomes treated in culture with 5-bromodeoxyuridine, and by Cohen and Shaw (1964) on the effect of mytomycin C on human chromosomes.

We believe that our observations should be continued and extended both from the point of view of qualitative changes,

Tab. II

PLATES BY CHROMOSOME NUMBER

Groups		Hypoploidy				Euploidy	Hyperploidy		Total
Classes		42	43	44	45	46	47	48	
Experimental sample	Frequencies of group	0.218				0.728	0.008		1.000
	Frequencies of class	0.021	0.021	0.046	0.130	0.728	0.046	0.008	1.000
Control sample	Frequencies of group	0.101				0.868	0.031		1.000
	Frequencies of class	0.000	0.003	0.034	0.064	0.868	0.031	0.000	1.000
Expected values*	Frequencies of group	0.10				0.850	0.05		1.000
	Frequencies of class	0.003	0.007	0.030	0.060	0.850	0.040	0.010	1.000

* According to Bartalos and Baramki (1967)

and from the point of view of quali-quantitative changes with age, with different products and with dosage.

As a conclusion to this introductory note we believe we can state that protracted use of the pill seems to induce a pattern of nuclear damage in circulating lymphocytes, which can be identified in the respective karyograms through an increase of polyploidy and a shift to the left (towards the hypoploid classes) in the distribution of diploid plates.

Yet it must be said that these findings are not specific, not being found in all women who take the pill. But it is well known that, clinically, different women may have different alterations and even physiologically different women may react in different ways in connection with the use of the pill; this finding led Gedda to formulate the general hypothesis that the pill may favour the expression of latent genetic defects, which of course are different in different individuals. This perspective could also explain the epidemiological differences of the side effects as a result of the genetic estate of each population.

If this should happen also at the level of the chromosomes, we could assume that the subjects revealing the karyologic transformations we reported are endowed with more fragile genomes. The concept that "nonrandomness" of chromosomal abnormalities may underline the existence of underlying genetic factors is shared by Hecht and coworkers (1964).

References

Bartalos M., Baramki T.A. (1967). Medical cytogenetics. The William & Wilkins Co., Baltimore.

Cohen M.M., Shaw M.W. (1964). Effects of mitomycin C on human chromosomes. J. Cell Biol., 23:386.

Carr D.H. (1969). Chromosomal abnormalities in clinical medicine. In A.G. Steinberg and A.G. Bearn: Progress in Medical Genetics. Vol. VI. Grune & Stratton, New York.

Hecht F.H., Bryant J.S., Gruber D., Townes P.L. (1964). The nonrandomness of chromosomal abnormalities. Association of trisomy 18 and Down's syndrome. New Eng. J. Med., 271:1081.

Hsu T.C., Somers C.E. (1961). Effect of 5-bromodeoxyuridine on mammalian chromosomes. Proc. Nat. Acad. Sci. USA, 47:396.

AMENORRHEA FOLLOWING ORAL CONTRACEPTIVES

Mirjam Furuhjelm, M D

The Department of Obstetrics and Gynecology, Karolinska Institutet, Sabbatsberg Hospital, Stockholm

Two important centres in the hypothalamus are believed to be concerned in the control of pituitary function. A centre in the medial part of the hypothalamus is believed responsible for a constant basal production of gonadotrophins maintaining basal ovarian activity. Another centre in the preoptic area of the anterior hypothalamus is thought responsible for surges of gonadotrophin release to maturation of the follicle and to ovulation. Organic lesions and probably also repression of the hypothalamus lead to disturbances at two levels of severity. In the milder form preoptic activity is abolished and surges of gonadotrophins do not occur, basal production however remains. Oestrogen production is at comparatively normal levels and menstruation may occur though cycles are of course anovulatory. In the more severe form the centre governing basal production of gonadotrophin is also involved. Gonadotrophin and oestrogen production falls and complete amenorrhea develops, (Ferriman 3).

From experiments on animals it is known that administration of oestrogen and progesteron inhibits the neuronal activity of the hypothalamus. Administration of these substances to human beeings dimishes the release of FSH and LH and thus inhibits ovulation. The excretion of FSH and LH in the urine is low and there is no ovulation peak of LH. Oral contraceptive suppression of FSH and LH is maximal after three treatment cycles. This suppression is maintained while the patient continues use of the agent. A longstanding medication of substances which diminish or inhibit the gonadotrophic releasing factors may also have a repressive effect on other releasing factors in the hypothalamus.

It has been beleived that after inhibition of hypothalamus a rebound effect follows after cessation of the treatment. Instead several reports (Whitelaw et al 8, Shearman 6,7, Halbert et al 4), have drawn attention to the fact that endocrine disturbances take place after the administration of contraceptive pills.

There is a higher frequency of amenorrhea and anovulatory cycles after the discontinuance of oral contraceptives than generally is believed. The disturbance might be related to a failure in restoration of the hypothalamic cycle of gonadotrophin releasing factors. The patients respond with ovulation after treatment with chlomiphene or gonadotrophic hormones.

Material: At the Hormone Laboratory of the University Department of Obstetrics and Gynaecology, Sabbatsberg Hospital, Stockholm, we have 1967-1969 determined the hormonal excretion in the urine of more than 100 women with longstanding amenorrhea after use of oral contraceptives. 30 patients have been amenorric for more than a year and their data are collected in Table I.

Almost all of the patients have a very low excretion of steroids. The repression of the hypothalamic releasing factors includes not only the gonadotropic hormones but also, probably, the ACTH releasing factor since the values for the excretion of 17-ketosteroids and 17-ketogene steroids are low. The patients have symptoms which may depend on a hypothalamic repression they are tired, thin, cold and complain of loss of libido.

Discussion: The low excretion of gonadotrophins and steroids in the urine together with other symptoms speak in favour of a repression of the hypothalamic releasing factors as beeing the cause of the amenorrhea. It may be possible that besides disturbances in the function of the ovaries have developed thus preventing the ripening of the follicles. We have inspected the ovaries at laparoscopy or laparotomy in all patients with amenorrhea more than for one year. In all cases we observed small inactive ovaries without any follicles. In 15 of the patients the excretion of dehydroepiandrosterone (DHA), androsterone (A), and etiocholanolone (E) was determined. In most cases no measurable amounts of DHA could be detected. Table II.

The follicles of the ovaries produce almost all of the ovarian DHA. The lack of production of DHA in cases with longstanding amenorrhea is therefore in good agreement with the finding at laparoscopy of absence of follicles in the ovaries. According to Mauvais-Jarvis (5) lynestrol and other oral contraceptives may have an influence of the activity of the ovarian enzymes. A change in the activity of 17-β -hydroxysteroid dehydrogenase may develop and hence disturbances in the synthesis of steroid hormones appear. The tranformation of pregnenolone to DHA is also inhibited. Appelgren (2) has recently shown that medroxy-progesteron and ethinylestradiol inhibits the enzymatic activity which in the ovaries transforms pregnenolone to pregesterone. According to Aakvaag (1) chlormadinon inhibits the same enzym system. They are consequently facts which

__Table 1__. Excretion of gonadotrophins and steroids in cases with longstanding amenorrhea.

Age	Para	Amenorrhea, years	G MU/24 hrs	E_1 E_2 /24 hrs	P mg/24 hrs	17-KS mg/24 hrs	17-KGS mg/24 hrs
19	0	1,0	< 10	10	0,2	5,5	8,3
20	0	1,0	< 10	1	0,3	10,6	6,8
20	0	2,0	< 10	3	0,9	6,7	-
21	I	1,3	< 10	5	0,5	6,1	-
21	0	1,0	< 10	2	0,2	5,7	3,4
21	0	1,0	< 10	5	0,7	10,5	7,6
22	0	1,0	< 10	3	-	6,9	9,0
22	0	1,5	< 10	5	-	7,9	4,2
22	0	1,0	-	3	1,5	13,7	-
22	0	1,5	< 10	1,5	0,5	5,8	3,9
23	0	1,5	>10<40	2	-	11,7	9,3
24	0	1,5	< 10	1	0,5	7,6	6,4
25	0	2,0	< 10	2	-	5,7	9,1
25	I	2,0	< 10	0	0,1	1,7	1,7
25	0	2,0	< 40	4	0,5	6,4	4,1
25	0	1,5	< 10	1	0,1	3,6	4,4
25	0	1,5	< 10	2	0,5	8,4	5,9
25	0	1,0	< 10	14	-	-	-
26	0	1,0	< 40	10	-	5,7	-
27	0	1,0	< 10	1	0,2	3,4	4,1
28	I	1,0	>10<20	5	-	7,0	3,2
29	0	1,0	-	-	-	3,9	2,7
29	0	1,0	< 10	7	0,1	-	-
29	0	1,0	< 10	6	-	3,4	4,6
29	0	1,5	-	3	-	3,2	2,9
30	II	2,0	< 10	2	0,3	3,5	6,0
31	0	1,5	< 10	2	0,3	5,9	3,5
31	II	1,0	< 10	1	0,4	4,9	5,3
33	III	3,0	< 10	4	0,3	5,7	4,5
36	II	1,0	-	3	0,5	3,4	4,9

Table II. Excretion of frectionated 17-ketosteroids.

Age	DHA mg/24hrs	%	A mg/24hrs	%	E mg/24hrs	%	11-Deoxy-17-KS mg/24 hrs
19	-	-	33,2	72,7	1,2	27,1	4,4
21	-	-	1,2	50,0	1,2	56,0	2,4
21	1,1	21,2	3,5	67,5	0,6	11,8	5,2
22	0,1	3,1	1,9	59,0	1,2	37,2	3,2
24	-	-	2,8	63,0	1,4	37,0	4,2
25	-	-	0,4	50,0	0,4	50,0	0,8
25	-	-	1,3	61,6	0,8	38,1	2,1
25	-	-	2,8	68,0	1,3	32,0	4,1
25	0,4	22,2	0,9	50,0	0,5	27,7	1,8
28	-	-	0,9	69,4	0,4	30,5	1,3
29	2,6	22,0	5,5	46,6	3,7	21,4	12,3
29	-	-	1,2	52,0	1,2	48,0	3,5
31	-	-	2,1	58,0	1,5	42,0	3,5
31	0,1	.2,8	1,1	35,5	1,6	51,7	3,1
35	-	-	0,7	46,6	0,8	53,0	1,5

speak in favour of the theory that oral contraceptives intervene as well with the function of the hypothalamus as with the function of the ovary itself.

Today it is not known how long the repression on the hypothalamus, depending on the action of oral contraceptives is going to persist. Neither do we know how to treat these patients with longstanding amenorrhea in order to get them to menstruate regularly. We must consider, that a longstanding treatment with ovulation inhibitors interferes with the normal hormonal balance of the woman. Recent works have clarified the mechanism of action of steroid hormones. They intervene in the synthesis of RNA and act as gene activators. If they also act upon the germ cells is not yet known.

REFERENCES

1. Aakvaag, A.: Effekten av clormadinon på konversjonen av pregnenolon til progesteron i homogenat av ovarier fra gris. Nordisk Fertilitetsklubbs 10 möte, Copenhagen 1969.

2. Appelgren, L.E.: Interference with progesterone synthesis of some antifertility compounds as shown by histochemistry. Congress Proceedings Second European Congress on Sterility, Dubrovnik 1969. Moragni Ed Ezianni Sientifiche Roma (in press).

3. Ferriman, D.: Anovulatory fertility. Medical gyn and sociol 4 2, 1969.

4. Halbert, D.R. and Christian, C.D.: Amenorrhea following oral contraceptives. Obstet and Gyn 34 161, 1969.

5. Mauvais-Jarvis, P.: Métabolisme de la testostérone et du sulfate de déhydroépiandrostérone radioactifs, chez des sujets traites par le lynoestrénol, les oestrogènes de synthèse et leur association. Acta Endocrin (Kbh) 57:37, 1966.

6. Shearman, R.P.: Amenorrhea after treatment with oral contraceptives. Lancet 2/7473:1110, 1966.

7. Shearman, R.P.: The investigation on treatment of amenorrhea developing after treatment with oral contraceptives. Internat J of Fertility 13:321, 1968.

8. Whitelaw, M.J., Nola, V.F, and Kalman, G.F.: Irregular menses, amenorrhea and infertility following synthetic progestational agents. J.A.M.A. 1958:720, 1966.

THE INFLUENCE OF ORAL CONTRACEPTIVES ON CARBOHYDRATE AND LIPID METABOLISM

Nils-Olov Lunell and Bengt Persson

Departments of Obstetrics and Gynecology and Pediatrics, Sabbatsberg and S:t Göran Hospitals, Karolinska Institute, Stockholm

Since 1964 when Gershberg et al. (14) published their report about impaired glucose tolerance (GT) in women taking oral contraceptives (OC), numerous investigations on this aspect have appeared. It is difficult to evaluate and summarize all results adequately because different types of steroids in different dosages have been administered to different types of populations for different lengths of time; furthermore the effects of these therapies have been measured with different types of parameters.

Most studies have dealt with combination types of drugs, and oral glucose tolerance tests (OGTT) have been performed. There are six such studies having control tests before OC were administered. Four of them (11, 27, 38, 44) showed no change on OC while two (29, 43) showed a deterioration of between 11 - 67% (Table 1). Results obtained without control studies have usually shown abnormal GT.

When intravenous glucose tolerance tests (IVGTT) were used to evaluate GT, no impairment was found in seven studies out of ten; in three the frequency of abnormal tests varied from 9 - 20% (Table 2). However, in four (4, 13, 38, 44) out of eight tests blood glucose levels were significantly higher than pretreatment levels.

If a corresponding analysis of results obtained during sequential therapy is undertaken, diverging opinions are found. Out of four studies, two (33, 44) showed normal tests while in two (25, 30) abnormal results were obtained in 17 and 45% of cases (Table 3). However, even in the investigation of Yen and Vela (44), where

TABLE 1

Influence of combination type of OC on OGTT[x]

Authors		Drug Estrogen	Progestogen	No. of women	% abnormal results
Danowski et al.	1968 (11)	M	Norethynodrel	17	0
Pi-Sunyer & Oster	1968 (27)	M	Norethynodrel	20	0
		M	Norethisterone	19	67[xx]
Puchulu et al.	1967 (29)	E	Norethisterone	24	11[xx]
Taylor & Kass	1968 (38)	M	Norethynodrel	21	0
Wynn & Doar	1969 (43)		Varying	66	17
Yen & Vela	1968 (44)	M	Ethynodiol	11	0

M = mestranol E = ethinylestradiol

[x] Only studies with pretreatment controls

[xx] OGTT and cortison GTT

TABLE 2

Influence of combination type of OC on OGTT[x]

Authors			Drug Estrogen	Progestogen	No. of women	% abnormal results
			M	Lynestrenol	8	0
Botterman et al.	1967	(4)	M	Chlormadinone	10	0
Clinch et al.	1969	(10)	M	Norethisterone	33	0
Frerichs et al.	1967	(13)	E	Norethisterone	17	0
			M	Ethynodiol	12	0
Posner et al.	1967	(28)	M	Norethynodrel	64	9
Spellacy et al.	1967	(31)	M	Norethynodrel	32	0
Starup et al.	1968	(35)	M	Megestrol	27	0
Waldhäusl et al.	1969	(39)	M	Lynestrenol	9	20
Wynn & Doar	1969	(43)		Varying	66	9
Yen & Vela	1968	(44)	M	Ethynodiol	11	0

M = mestranol E = ethinylestradiol

[x] Only studies with pretreatment controls

TABLE 3

Influence of sequential type of OC on OGTT or IVGTT[x]

Authors		Drug Estrogen	Progestogen	No. of women	Test	% abnormal results
Spellacy et al.	1969 (33)	M	Chlormadinone	28	IVGTT	0
di Paola et al.	1968 (25)	M	Chlormadinone	43	OGTT	45[xx]
Pyörälä et al.	1967 (30)	E	Megestrol	6	IVGTT	17
Yen & Vela	1968 (44)	E	Dimethisterone	12	IVGTT OGTT	0

M = mestranol   E = ethinylestradiol

[x] Only studies with pretreatment controls

[xx] OGTT and cortison GTT

no pathological tolerances were recorded, blood glucose values during IVGTT were significantly elevated during OC.

One explanation of the diverging results might be the populations studied. Wynn and Doar (43), who found deterioration of GT in a high frequency, had a material where 48% of the women had a diabetic heredity. Spellacy et al. 1969 (33) found a significant association between elevations of glucose and insulin levels and family history of diabetes. Gerasi and Luft, 1967 (9) showed that around 15% of a normal population had a delayed insulin release to a glucose stimulus and postulated that these were prediabetic. It might be that such people would be more prone to develop impaired GT on OC.

We have thought it would be of interest to investigate individuals with a high risk for developing diabetes. Women having had gestational diabetes, i.e. a diabetic GT during pregnancy which normalizes after delivery, are according to the classification of WHO (40) latent diabetics. If OC would impair GR, the effects should be most prominent in latent diabetics.

We have thus studied five women with gestational diabetes. Repeated IVGTT were performed. The intravenous GT was measured as the disappearance rate of blood glucose expressed with a k-value. A k-value below 1.0 is pathologic (22). One woman (M 22, Fig. 1), having a diabetic k-value of 0.7 during pregnancy and a normal GT with k 1.2 after delivery, showed during OC a decrease to 0.7. The medication was stopped and an IVGTT then gave a k-value of 1.3. However, 4 months later she had developed clinical diabetes. Another woman (M 11, Fig. 2) also had a k-value during the last trimester of 0.7 which after delivery increased to 1.9. After 3 months of OC the GT was pathological with k = 0.8. The administration was stopped and another IVGTT gave a k of 1.2. She was again given OC. At first the GT improved and 2 IVGTT showed k-values of 1.9. However, after another 4 months the k-value decreased to 1.1.

Three more women with gestational diabetes showed principally the same pattern with deterioration of GT during OC. They are shown in Fig. 3, 4 and 5. Thus it seems that OC could impair GT in women with a diabetic trait. Our results are in good agreement with the recent report by Szabo et al. 1970 (37).

Growth hormone (GH) has been found to increase after estrogen administration (12, 23) and also to induce insulin release (1, 3, 8). Increased GH (34, 44) and insulin (32, 39, 43, 44) levels have been measured during OC. However, unchanged values of insulin have also been found (4, 33). GH is diabetogenic and decreases GT (18, 21). It might thus be that estrogen stimulates increased GH secre-

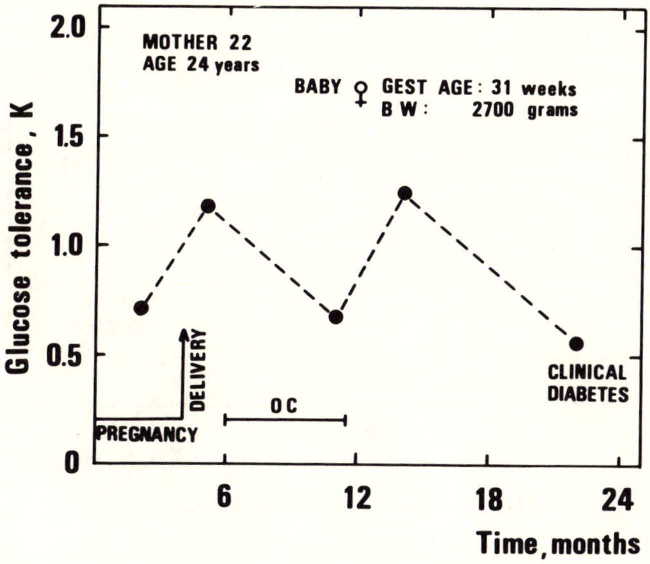

OC = MESTRANOL 0.1 mg + NORETHISTERON 2 mg

Fig. 1

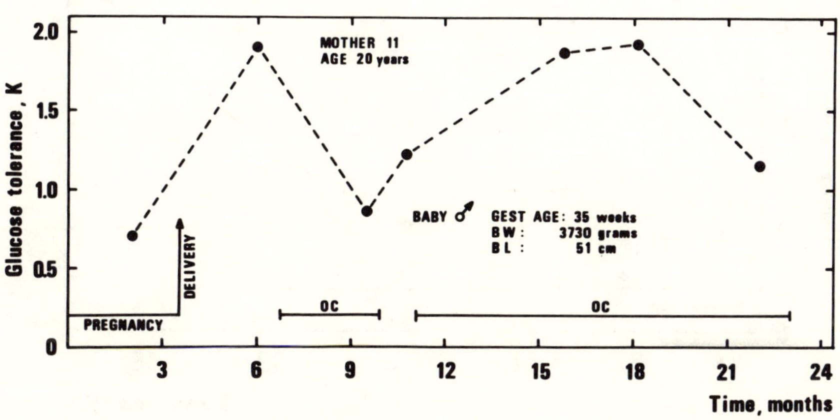

OC = MESTRANOL 0.075 mg + LYNESTRENOL 2.5 mg

Fig. 2

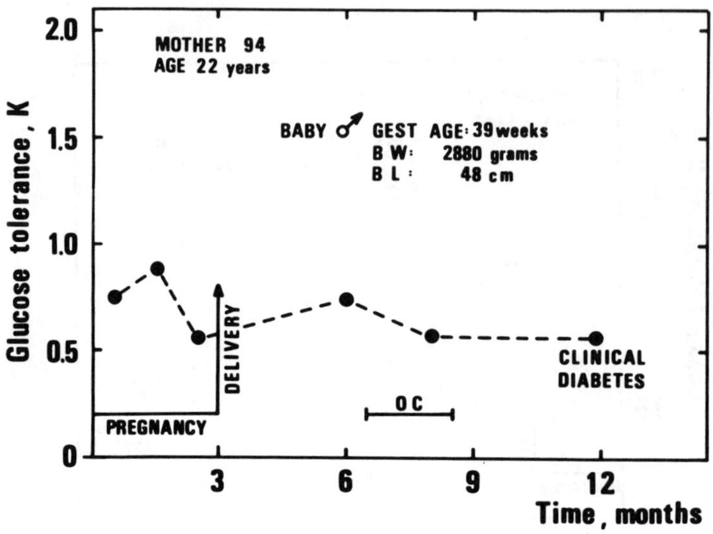

OC = MESTRANOL 0.1 mg + NORETHISTERON 1 mg

Fig. 3

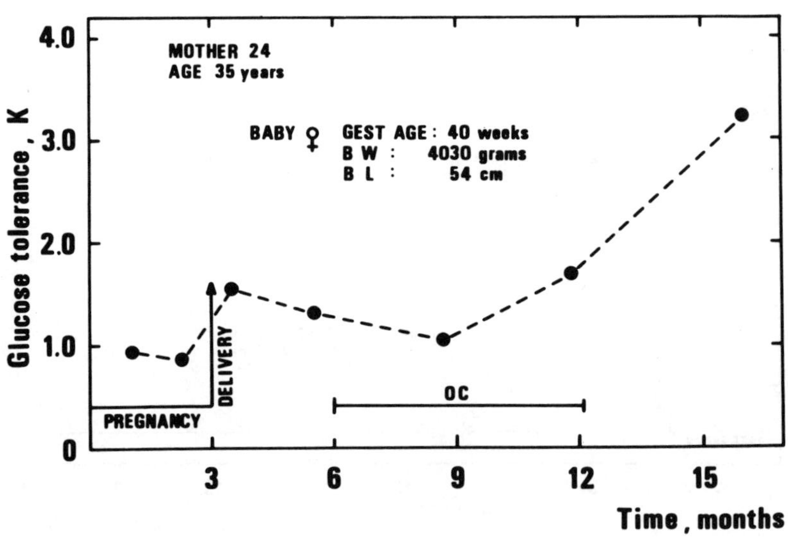

OC = ETHINYLESTRADIOL 0.05 mg + NORETHISTERON 4 mg

Fig. 4

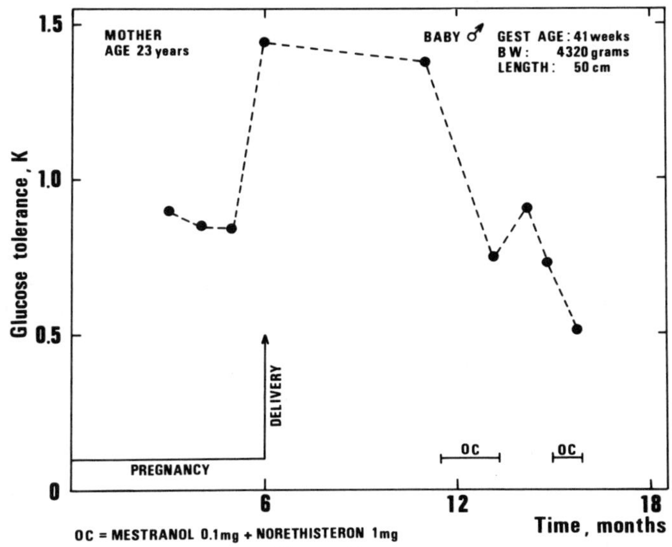

Fig. 5

tion which in normal individuals will induce a compensatory hyperinsulinism overcoming the diabetogenic effect of GH. In a subject with impaired pancreatic β -cell function, the compensatory mechanism would fail and the end-result would be an abnormal GT.

Lipid metabolism

The alterations induced by OC on lipid metabolism are more consistent than those on carbohydrate metabolism. Increased levels of serum triglycerids, very low density lipoproteins and high density lipoproteins have been found (2, 5, 17, 20, 41, 42, 45). Usually no changes in serum cholesterol have been recorded (32, 43, 44), but higher values have also been measured (2, 20, 42). Although the cause of the changes in lipid metabolism remains unknown a number of observations are pertinent in this respect. Post-heparin lipoportein lipase activity is decreased during OC (17), which if reflecting a decreased tissue lipoprotein lipase activity, could result in a decreased capacity to remove triglycerides from the blood. The elevation of plasma triglyceride concentrations could also be explained by increased inflow of triglycerides to the blood stream. This explanation might be supported by the finding of elevated plasma free fatty acids (FFA) concentrations (42). Thus OC therapy may cause increased mobilization of FFA from adipose tissue, increased hepatic uptake of FFA, and increased hepatic synthesis and release of triglycerides. Plasma glycerol concentration, an indirect measure of lipolysis, have not yet been studied. OC might also alter the specific pathways of hepatic FFA metabolism by diverting a greater than normal amount of FFA from ketogenesis to triglyceride synthesis. If this were true the normal female pattern of low blood triglycerides and high ketone bodies would be altered to one of high triglycerides and low ketone bodies, the pattern observed in normal males (19).

Ketone bodies have not yet been determined in women during OC therapy. The manifestations of ischaemic cardiovascular disease appear to be related to elevations of triglycerides (6, 7, 24). OC therapy not only raise triglyceride concentrations but may also eliminate the possible protective female pattern of low triglycerides and high ketone bodies (19). The possibility therefore exists that the use of OC for long periods of time might deprive the woman of her privilege to be preserved from ischaemic cardiovascular disease before the menopause.

In view of the described changes in carbohydrate and lipid metabolism special caution is in order when OC are given to patients with different stages of diabetes.

REFERENCES

1. Altszuler, N., Steele, R., Rathgeb, I. and de Bodo, R.C.: Growth Hormone Proc. 1st Int. Symp. Exp. Med. Congr. Ser. No. 158, p. 309, 1968.

2. Aurell, M., Cramer, K. and Rybo, G.: Lancet 1, 291, 1966.

3. Beck, P., Schalch, D.S., Parker, M.L., Kipnis, D.M. and Daughaday, W.D.: J. Lab. Clin. Med. 66, 366, 1965.

4. Botterman, P., Dieterle, P., Hochheuser, W., Horn, H., Kopetz, K., Schleypen, K., Schwarz, K. and Scriba, P.C.: Münch. Med. Wschr. 13, 685, 1967.

5. Brody, S., Verstell, J., Nilsson, L. and Svanborg, A.: Acta Med. Scand. 183, 1, 1968.

6. Brown, D.F., Kinch, S.H. and Doyle, J.T.: New Engl. J. Med. 273, 947, 1965.

7. Carlsson, L.: Acta Med. Scand. 167, 399, 1960.

8. Cerasi, E. and Luft, R.: Lancet 2, 769, 1964.

9. Cerasi, E. and Luft, R.: Acta Endocr. (Kbh) 55, 278, 1967.

10. Clinch, J., Turnball, A.C. and Khosla, T.: Lancet 1, 857, 1969.

11. Danowski, T.S., Sabeh, G., Alley, R.A., Robbins, T.J., Tsai, C.T. and Sekaran, K.: Clin. Pharmacol. Therap. 9, 223, 1968.

12. Frantz, A.G. and Rabkin, M.T.: J. Clin. Endocr. 25, 1470, 1965.

13. Frerichs, H., Grote, S., Severidt, E. and Creutzfeldt, W.: 12. Symp. Deutsch. Ges. Endokrinologie, 1966. Springer Verlag, Berlin, p. 123, 1967.

14. Gershberg, H., Javier, Z. and Hulse, M.: Diabetes 13, 378, 1964.

15. Gershberg, H., Hulse, M. and Javier, Z.: Obstet. Gynec. 31, 186, 1968.

16. Gershberg, H., Zorilla, E., Hernandez, A. and Hulse, M.: Obstet. Gynec. 33, 383, 1969.

17. Hazzard, W.R., Spiger, M.J., Bagdade, J.D. and Bierman, E.: New Engl. J. Med. 280, 471, 1969.

18. Ikkos, D. and Luft, R.: Acta Endocr. (Kbh) 39, 567, 1962.

19. Jenkins, D.J.A., Welbdrn, T.A. and Goff, D.V.: Lancet 1, 865, 1970.

20. Larsson-Cohn, U., Berlin, R. and Vikrot, O.: Acta Endocr. (Kbh) 63, 717, 1970.

21. Luft, R., Cerasi, E. and Hamberger, C.-A.: Acta Endocr. (Kbh) 56, 593, 1967.

22. Lunell, N.O.: Acta Obstet, Gynec. Scand. 45, Suppl. 4, 1966.

23. Merimee, T.J., Rabinowitz, D., Riggs, R., Burgess, J.A., Rimoin, D.L. and Mc Kusick, V.A.: New Engl. J. Med. 276, 434, 1967.

24. Ostrander, L.D., Neff, B.J., Block, W.D., Francis, T. and Epstein, F.H.: Ann. Intern. Med. 67, 34, 1967.

25. di Paola, G., Puchulu, F., Robin, M., Nicholson, R. and Marti, M.: Am. J. Obst. Gynec. 101, 206, 1968.

26. Pincus, G.: The Control of Fertility, Academic Press, New York, p. 276, 1965.

27. Pi-Sunyer, F.X. and Oster, S.: Obstet. Gynec. 31, 482, 1968.

28. Posner, N.A., Silverstone, F.A., Pomerance, W. and Baumgold, D.: Obstet. Gynec. 29, 79, 1967.

29. Puchulu, F., di Paola, G., Marti, M.L., Robin, M., Nicholson, R. and Groppar, S.: VI Congr. Int. Diab. Fed. 1967, Exerpta Med. Int. Congr. Ser. No, 140, p. 122.

30. Pyörälä, K., Pyörälä, T. and Lampinen, V.: Lancet 2, 776, 1967.

31. Spellacy, W.N., Carlsson, K.L. and Birk, S.A.: Diabetes 16, 590, 1967.

32. Spellacy, W.N., Carlsson, K.L., Birk, S. and Schade, S.L.: Metabolism 17, 496, 1968.

33. Spellacy, W.N., Buchi, W.C. and Bendel, R.P.: Obstet. Gynec. 33, 800, 1969.

34. Spellacy, W.N., Buchi, W.C. and Bendel, R.P.: Int. J. Fertil. 14, 51, 1969.

35. Starup, J., Date, J. and Deckert, T.: Acta Endocr. (Kbh) 58, 537, 1968.

36. Svanborg, A. and Vikrot, O.: Acta Med. Scand. 179, 615, 1966.

37. Szasbo, A.J., Cole, H.S. and Grimaldi, R.D.: New. Engl. J. Med. 282, 646, 1970.

38. Taylor, M.B. and Kass, M.B.: Am. J. Obstet. Gynec. 102, 1035, 1968.

39. Waldhäusl, W., Beringer, A. and Schneider, W.: Geburtsh. Frauenheilk. 29, 611, 1969.

40. WHO Techn. Report Ser. No. 310, 1965.

41. Wynn, V., Doar, J.W.H. and Mills, G.L.: Lancet 1, 720, 1966.

42. Wynn, V., Doar, J.W.H., Mills, G.L. and Stokes, T.: Lancet 2, 756, 1969.

43. Wynn, V. and Doar, J.W.H.: Lancet 2, 761, 1969.

44. Yen, S.S.C. and Vela, P.: J. Clin. Endocr. 28, 1564, 1968.

45. Zorilla, E., Hulse, M., Hernandez, A. and Gershberg, H.: J. Clin. Endocr. 28, 1793, 1968.

THROMBOEMBOLISM AND ORAL CONTRACEPTIVES

Barbro Westerholm

Department of Pharmacology, Karolinska Institute

Stockholm

An association between thromboembolic disease and the use of oral contraceptives was first suspected when Jordan (1961) described pulmonary embolism in a woman using Enavid (norethynodrel and ethinyloestradiol). Since then numerous accounts have been published in the medical and lay press of individual or small groups of patients who have developed thromboembolic disorders while taking oral contraceptives.

In order to prove whether or not a cause-effekt relationship existed between oral contraceptives and thromboembolism considerable attention has been focused on coagulation mechanisms in women taking the pill and on mortality and morbidity statistics for thromboembolism in women in the fertile age group.

Effects on Blood Coagulation

There have been a multitude of reports, many conflicting, concerned with changes of coagulation factors in oral contraceptive users. Seemingly all parameters have been investigated; Table 1 is an attempt to summarize the findings.

Most of the studies demonstrate an increase in Factors II, VII, and X, and Factor VII-X (P and P) complex with the use of oral contraceptives (for ref. see Ygge *et al.* 1969). There is no decrease in fibrinolytic activity as seen in pregnancy. Most authors agree that these elevations in coagulation factors do not predispose to venous thrombosis.

Table 1

Factor II	+
Factor VII	+
Factor VIII	+ or 0
Factor X	+ or 0
Fibrinogen	+ or 0
Plasminogen	+ or 0
Antiplasmin activity	+
Fibrinolytic activity	+ or 0
Antifibrinolytic activity	+
Platelet adhesiveness	+ or -
Platelet number	+

Changes in blood coagulation in women receiving oral contraceptives.

An interesting finding, however, is that some factors involved in blood clotting and fibrinolysis increase during treatment with oral contraceptives (Ygge et al. 1969). Approximately in the middle of the treatment period, there is a maximal concentration of fibrinogen, plasminogen and spontaneous fibrinolytic activity which then successively decreases toward the end of the tablet period. It has been suggested that this leads to a periodical imbalance in the hemostatic mechanism. If, during this period, other factors are present which predispose to thrombosis the changes in blood coagulation and fibrinolysis may be the factor which causes the thrombosis.

Mortality and Morbidity Statistics

It was not until 1967 that enough statistical evidence was collected to show a correlation between the use of oral contraceptives and death from pulmonary embolism or cerebral thrombosis in the absence of predisposing conditions. The mortality from these two diseases attributable to the use of oral contraceptives by healthy women was estimated at 1.3 per 100,000 users aged 20 - 34 and 3.4 per 100,000 users aged 35 - 44 per annum (Inman and Vessey 1968). In Britain the risk of hospital admission for venous thromboembolism is estimated to be one in 2,000 for women using oral contraceptives compared with one in 20,000 for women not using them (Vessey and Doll 1968).

The British findings have later been confirmed in two American studies (Sartwell et al. 1969, Markush et al. 1969).

Swedish Figures

During the period October 1965 - December 1969 the Swedish Adverse Drug Reaction Committee has received reports on 131 women with no other known diseases who had been using the pill when hospitalized because of deep venous thrombosis. Fifty per cent of the women were 25 years old or younger. They had been using oral contraceptives for periods varying from less than one month up to more than three years when taken ill. Fifty per cent fell ill within the first six months of medication.

As in other countries, the possibility that oral contraceptives cause thromboembolism led to investigations on mortality and morbidity in thromboembolic disorders among women of the fertile age group.

It could be shown that the mortality in thromboembolism in women aged 15 - 44 years had not changed after 1964 when the oral contraceptives were introduced on to the Swedish market (Table 2). This is probably due to the fact that death due to thromboembolism following the use of oral contraceptives is rare, especially in the age group in which oral contraceptives are most commonly used in Sweden, i.e. below the age of 30 (Westerholm 1969).

Table 2

Year	No. of deaths per 100,000	
	men	women
1961	6.2	1.1
1962	5.3	1.4
1963	7.7	1.5
1964	7.9	1.9
1965	8.8	2.6
1966	7.6	2.3
1967	7.5	2.4
1968	5.8	2.1

Mortality in thromboembolism in Sweden 1961 - 1968
Age group 15 - 44 years.

Table 3

Year	No. of cases: Oral contraceptive users	Not on oral contraceptives	Oral contraceptive medication unknown	Total	Per cent oral contraceptive users in population[1]
1964	1	5	1	7	1
1965	8	2	3	13	7
1966	6	4	4	14	10
1967	17	10	1	28	16
1968	17	10		27	20

Morbidity in thromboembolism in Uppsala region, Malmö and a Stockholm area (1.5 million inhabitants) 1964 - 1968. Women aged 15 - 44 years hospitalized because of thromboembolism without predisposing illnesses.

1) Calculated from the sales on the assumption that all the tablets sold were consumed and that the women used the oral contraceptives for thirteen cycles per year.

The morbidity in thromboembolism has been studied in an area covering about one fifth (1.5 million) of the Swedish population. It was found that thromboembolism without predisposing diseases is rare in women aged 15 - 44 years. There has been an increase since the oral contraceptives were introduced in 1964 both in women using the pill and in women using other contraceptive methods (Table 3). However, statistical evaluation shows that there is a significantly higher increase in morbidity in thromboembolism in women on the pill compared with non users ($p < 0.001$).

Predisposing Factors

It is recommended that women with a previous history of thrombosis should not use oral contraceptives, although at present there is no conclusive evidence that they are at a higher risk. There is, however, one study which points out that certain women are more likely to develop thrombosis than others.

A prospective drug surveillance programme in Boston revealed a deficit of patients with blood group O among those who received anticoagulants for venous thromboembolism. This observation led to a cooperative study in the USA, Sweden and the UK to obtain the blood types of young white women developing venous thromboembolism while taking oral contraceptives, during pregnancy or the puerperium, or at other times. There was a deficit of patients with blood type O in all the groups studied. This deficit was particularly prominent when the thromboembolism was associated with the use of oral contraceptives or with pregnancy (Jick et al. 1969).

Steroidal Content of Oral Contraceptives

In the UK the Committee on Safety of Drugs found in 1967 that proportionally more reports on thromboembolism were obtained for oral contraceptives containing mestranol than could be expected from the sales. A closer analysis revealed that it was the size of the oestrogen dose which was important, rather than which of the two oestrogens - mestranol or ethinyloestradiol - that was given. This finding led to a cooperative study by the drug safety committees in the UK, Sweden and Denmark (Inman et al. 1970). For each country the number of reports of thromboembolism for each product was compared with the distribution that would have been expected from market research estimates of sales, assuming that all products carried the same risk.

The UK study showed a positive correlation between the dose of oestrogen and the risk of pulmonary embolism, deep vein thrombosis, cerebral thrombosis, and coronary thrombosis. A similar association was found for venous thrombosis and pulmonary embolism in Sweden and Denmark. Certain discrepancies in the data suggest that the dose of oestrogen may not be the only factor related to the risk of thromboembolism. There is some evidence that this might depend upon the progestogen used.

Conclusion

It has been proved that oral contraceptives can cause deep venous thrombosis, pulmonary embolism and cerebral thrombosis. Women belonging to blood group O are less likely to develop deep venous thrombosis and pulmonary embolism than women belonging to blood group A when using oral contraceptives. A positive correlation between the dose of oestrogen and the risk of thromboembolism has been found. There is evidence that the occurrence of thrombosis is also related to the type of progestogen used.

References

Brit. Med. J. 2:355-359, 1967.

Inman, W.H.W. and M.P. Vessey. Brit. Med. J. 2:193-199, 1968.

Inman, W.H.W., M.P. Vessey, B. Westerholm and A. Engelund. Brit. Med. J. 203-209, 1970.

Jick, H., D. Slone, B. Westerholm, W.H.W. Inman, M.P. Vessey, S. Shapiro, G.P. Lewis and J. Worcester. Lancet 1:539-542, 1969.

Jordan, W.M. Lancet 2:1146-1147, 1961.

Markush, R.E. and D.G. Seigel. Amer. J. Public Health 59:418 - 434, 1969.

Sartwell, P.E., A.T. Masi, F.G. Arthes, G.R. Greene and H.E. Smith. Amer. J. Epidemiol 90:365-380, 1969.

Vessey, M.P. and R. Doll. Brit. Med. J. 2:199-205, 1968.

Westerholm, B. Alcoholic Cirrhosis and Other Toxic Hepatopathias. Skandia International Symposia. Editors: A. Engel and T. Larsson. Nordiska Bokhandelns Förlag, Stockholm, 1970. In press.

Ygge, J., S. Brody, K. Korsan-Bengtsen, L. Nilsson. Amer. J. Obstet. Gynecol. 104:87-98, 1969.

POSSIBLE MECHANISM OF THE ANTI-FERTILITY EFFECT OF COPPER

Fred A. Kincl and Harry W. Rudel

The Population Council, The Rockefeller University, N. Y.

and Biological Concepts, Inc., New York, N. Y.

Zipper et al.(1969) reported that a small copper (or zinc) loop placed in the lower portion of one of the uterine horns in rabbits produced a decrease in the number of implantation sites without producing a significant alteration in the control uterus. Silver, tin and magnesium were not effective. Zipper and co-workers (1970) utilized this observation to design a copper-containing intrauterine device for the prevention of fertility in the human female and reported that clinical effectiveness was related to the amount of copper present (Table 1).

TABLE 1

CONTRACEPTIVE EFFECTS OF COPPER AFTER ONE YEAR OF USE

	IUCD	IUCD(a)	IUCD(b)
Pregnancy rate	18.3	4.9	0.9
Continuation rate	72.9	87.5	94.5
Closures:			
Expulsions	6.3	5.0	0.6
Bleeding	1.8	1.8	0.0
Pain	0.4	0.0	1.5
Months of use	4262	2516	1635

a 30 mm^2 of copper

b 120 mm^2 of copper

We have studied anti-fertility effectiveness of copper and other metals in rats and rodents and have undertaken experiments to define possible mechanism of action (Chang et al., 1970). In rats insertion of a small copper loop (1-3 mm) on day 3 after insemination completely prevented pregnancy in the experimental horn, but not in the control horn. When a copper loop of similar size was inserted on day 6, rather than day 3 after insemination, fertility was inhibited only partially. It was noted, however, that all fetuses in "copper treated" horns were apparently normal; no macroscopic evidence of skeletal or visceral malformation was evident.

Combined data from several experiments for the metals studied are summarized in Table 2. In this table effectiveness (fertility index) has been calculated by dividing the number of viable implants in the treated horn (X 100) by the number of viable implants in the control horn. Of the 10 metals studied only copper was found to consistently inhibit fertility in rats; no viable implants were seen in the group of 21 rats studied. The high activity of zinc, cobalt, cadmium, and lead is indicated by the low fertility indices.

TABLE 2

ANTI-FERTILITY EFFECTS OF VARIOUS METALLIC IUD'S IN RATS

Metal used	No. of rats	Average fertility index
None	14	101.6
Copper	21	0.0
Zinc	12	3.7
Platinum	15	20.0
Silver	16	21.1
Gold	13	26.2
Nickel	5	7.1
Monel	5	8.3
Cobalt	5	3.0
Cadmium	5	3.2
Lead	5	3.4
Silver and Platinum	12	58.3

Monel metal was moderately active whereas platinum, gold, and silver were only marginally active. Ovarian function was not inhibited by the presence of the metals in the uterus. In all the groups the average number of corpora lutea found in the ovary of the treated horn was about the same as found on the control side, and in untreated control rats.

In hamsters, copper was tested for contraceptive effectiveness in a group of 9 females. One female was found with one viable fetus and no resorptions in the treated horn. In the control horn the average number of viable implants was about 7.

Metallic copper had a profound effect on the endometrial response to decidualization in intact, pseudopregnant rats. It caused a partial inhibition in this test and the average weight of traumatized horns was only about one-half that of the controls. Our observation is in agreement with that of Prager (1969) who reported that the synthesis of DNA and proteins in the rat endometrium was markedly reduced in the presence of copper or zinc. An additional insight into the possible mechanism of action was obtained by following the effect of metallic copper upon the development of morulae and blastocysts in rats (Table 3). On day 4 after insemination morulae (but not blastocysts) were found in the uterine lumen containing the copper loop. One day later blastocysts were seen only in control horns and were absent in the "treated" horns. Chang and Tatum (1970) reported that blastocysts transplanted from a "copper" horn to a control, pseudopregnant animal will implant.

TABLE 3

RECOVERY OF MORULAE AND BLASTOCYSTS FROM THE UTERUS OF RATS HAVING A COPPER IUD

No. of rats	Recovery day	Average No. recovered (range) Morulae Cu-H	Morulae C-H	Blastocysts Cu-H	Blastocysts C-H
8	4	1.5(0-3)	2.6(1-4)	0	0
8	5	0.5(0-2)	0.4(0-2)	0	2.6(1-4)

C-H, control horn
Cu-H, uterine horn with copper

The animal studies indicate that in the presence of copper there is a change in the intrauterine environment which may interfere with the development of the morulae into blastocysts. In addition, if blastocysts do develop, their implantation is inhibited.

The function of trace metals in living systems is to act mainly as biological catalysts. Copper is probably one of the most ubiquitous catalysts of all the metals and participates in a great variety of reactions (Frieden, 1962). It can react with a variety of amino acids and proteins, more strongly than any other metal ion, to form very stable chelates, and copper enzymes are essential in many oxydative reactions. In addition, it is a highly specific antagonist for a very similar metal from this group, zinc (van Reen, 1953). One of the most important zinc-containing enzymes of the female reproductive tract is carbonic anhydrase, an enzyme directly associated with a progestational response in many species. It is known, for example, that carbonic anhydrase activity is high during pregnancy in the sheep; its activity is present in the deciduoma of rats, and in rabbits the levels rise following progesterone stimulation. It is attractive to speculate that the anti-fertility of metallic copper in rodents, and possibly in humans, may be due to an interference with a metal-containing enzyme, such as the carbonic anhydrase.

REFERENCES

Chang C.C., and Tatum H.J.: Contraception 1 (1970) 265.

Chang C.C., Tatum H.J., and Kincl F.A.: Fertil. and Steril. 21 (1970) 274.

Frieden E. In: Horizons in Biochemistry, Kasha M. and Pullman B., Eds., Academic Press, New York (1962).

Prager R.: Fertil.and Steril. 20 (1969) 944.

Van Reen R.: Arch. Biochem. Biophys. 46 (1953) 337.

Zipper J., Medel M. and Prager R.: Amer. J. Obstet. Gynec. 105 (1969) 529.

Zipper J., Tatum H.J., Medel M., Pastene L. and Rivera M.: Presented at the 15th Nobel Symposium on Control of Human Fertility, May 27-29, 1970, Stockholm.

INTRA-UTERINE DEVICES: SIDE EFFECTS

Carl-Axel Ingemanson, M.D.

Obst. and Gynaec. Dept. Central Hospital

Eskilstuna, Sweden

Theoretically, IUD ought to be the almost perfect contraceptive. The method is cheap, insertion is simple, there is no patient error and the IUD can remain in the uterus as long as the woman wants. There are some factors limiting the usefulness of the method, however, with the IUDs available for common use at present. There is a low but significant pregnancy rate.

Most IUDs have a relatively high expulsion rate. The rate of expulsion is dependent on the shape, size and stiffness of the device. As a rule, a small or more flexible IUD is more easily expelled than a larger or more rigid one. Expulsion is more common among young women and women without children. Expulsion usually occurs during the menstrual period and the first months after insertion, and is uncommon after the first six months. Expulsion is an annoying failure of the method, but it also creates another problem. In many cases the expulsion is not noticed by the patient and an unwanted pregnancy occurs.

Intrauterine infection and salpingitis was earlier considered to be a common complication. In many cases these infections probably represented a venereal infection, an exacerbation of preexisting infection or an infected abortion; only in a few cases is it likely that the infection was caused by the IUD. Investigations have shown that bacterial contamination of the uterus occurs at the time of insertion but after one month all bacteria have been eliminated. Thus, the common histologic picture of chronic endometritis cannot be correlated with the presence of bacteria. In our opinion, genital infection is not a major problem if a careful examination is made with due precautions against infection and sterile handling and insertion of the device. If a salpingitis occurs, it can be treated with success with the IUD in situ in most cases.

Perforation of the uterus during insertion can occur. There seems to be a special risk with the Birnberg bow, for which the perforation rate is about 1 per 100 insertions even when inserted by experienced physicians. The perforation rate with the Lippes loop is much less about 1 per 2,000 insertions. Perforation occurs more frequently in a soft, recently gravid uterus and most of the IUD-perforations have occured 5-7 weeks post partum. Intestinal obstruction has been reported with closed devices such as the Birnberg bow, in which case a small bowel has become herniated through the device. No such complications have been reported with open devices such as Lippes loop, but as a rule it is advisable to remove an IUD inserted in the abdominal cavity.

Heavy and/or prolonged bleeding, sometimes associated with pain, occurs frequently during the first few months after insertion of intrauterine devices. Minor cramps just after the insertion are common, but pain alone is not common and dysmenorrhoea does not seem to be deteriorated by IUDs.

The rate of removals due to bleeding declines with increasing duration of use, but it continues to be the chief cause of method failure.

Table I shows the opinions of some of our IUD patients concerning their menstrual periods after the first 6 months following insertion of Lippes loop C and the Saf-T-coil.

Table I

Menses after the first six months following IUD insertion (515 pats Lippes loop C, 392 pats Saf-T-coil)

	As before IUD	More abundant	More prolonged	More abundant and prolonged
Lippes loop	31,6 %	30,4 %	15,6 %	22,4 %
Saf-T-coil	25,4 %	26,1 %	20,4 %	28,1 %

	Spotting between menses
Lippes loop	16,5 %
Saf-T-coil	25,1 %

Only about one-fourth of the women consider their menstruation to be the same as before insertion. The majority of the women report heavier and/or prolonged bleeding, and spotting in between the periods is common. In spite of this, only 5 % in both groups wanted to change to another contraceptive. The reason for increased bleeding among IUD-wearers is not quite clear, but it can partly be explained by alterations in the endometrium. The endometrial tissue is often oedematous and is sometimes hyperplastic, and pressure effects from the device can be seen. The endometrium adjacent to the device may become ulcerated and dilated blood vessels are common.
Another factor that seems to increase bleeding is calcium deposits on plastic IUDs.
(Fig. I)

Such calcium deposits are quite common and since calcium ions are a necessary part of the clotting mechanism, the device may possibly interfere with coagulation by local calcium depletion. It is also possible that calcium deposits cause increased bleeding through mechanical action.
The deposition of calcium was an early problem with prosthetic heart valves, which was solved by adding silicone to the plastic of the valves. Dr Lippes added 2 % gum silicone to the D-loop and a comparison of the one-year figures for the Silicone loop D and the usual D-loop showed a significantly lower rate of removals due to bleeding and/or pain with the Silicone loop.
A great number of IUDs have been designed in different shapes, sizes and materials. Among the best known are The Ota-ring, the Birnberg bow, the Hall-Stone ring, the Margulies spiral, the Saf-T-coil and the Lippes loop. From most points of view Lippes loop has proved to be the best device and is the IUD that is most widely used for the time being.
However, this device cannot claim to be an ideal IUD. The pregnancy and expulsion rate of removal for medical reasons

is too high.
New IUDs have been designed which seem to be superior to the devices used up to now as far as can be judged from preliminary trials.
Dr Lippes has developed a new loop - the Shell loop - a silicone-treated polypropylene loop (named after the Shell Oil Corporation, who produce the polypropylene). This loop is of the same size and shape as the D-loop but it is much more rigid. The pregnancy rate is very low - 0,5 - and the expulsion rate is also low with 3,5 % first expulsions, and the medical removal rate is lower than with the usual polyethylene loop D.

Table II

Net cumulative event rates, by type of termination, and total closure and continuation rates per 100 users, one year of use. (Life table method, Dr Tietze, Population Council)

Type of termination	Lippes Shell loop Rate	Lippes loop D Rate
Events		
Pregnancy	0,5	2,7
Expulsions:		
First	3,5	9,5
Later	0,3	3,2
Removals:		
Bleeding/pain	8,3	11,7
Other medical	3,7	3,5
Planning pregnancy	2,0	0,9
Other personal	2,2	2,2
Closure rate	16,9	22,6
Continuation rate	83,1	77,4
First insertions	1241	7553
Woman-months of use	9696	72046

We have used the Shell loop in trials at our clinic since May, 1969, and our preliminary experiences are very satisfactory. Another promising IUD is the Dalcon Shield device, which has a low pregnancy rate and a very low rate of expulsion and medical removals.
Tatum and Zipper have designed another new interesting IUD, the T-device with copper as a thread or a socket around the device. Copper has been found to possess potent anticonceptional properties, as can be seen in Table III, showing a comparison between a T-device without copper and a T-device with a copper thread. This device has a low expulsion rate and a very low medical removal rate.

Table III

Cumulative rates of events and closures per 100 users and one year of use for devices T-plain and T Cu-B (Life table method, Dr Tietze, Population Council)

Type of termination	T-plain Rate	T Cu-B Rate
Events		
Pregnancy	18,3	4,9
Expulsions:		
First	5,9	4,5
Later	0,4	0,5
Removals:		
Bleeding/pain	2,2	1,8
Other medical	1,0	2,0
Planning pregnancy	2,6	4,6
Other personal	1,3	0,4
Closure rate	26,9	12,1
Continuation rate	72,9	87,5
Nr of patients	605	279
Women months of use	4202	2516

In conclusion, it can be ascertained that there are considerable possibilities for development in the field of intrauterine contraception. In the near future we shall probably achieve highly effective protection against pregnancy with a minimum of side effects by using improved intrauterine devices, and we can thereby avoid the actual and potential hazards of systemic medication for birth control.

References

Basic and clinical aspects of intrauterine devices. WHO technical report series No 332, 1966, Geneva.

Davis, H.J. Am. J. Obstet. & Gynec. 106:455, 1970

Hagbard, L. and Ingemanson, C-A. Svenska Läkartidningen 66:3285, 1969

Hall, R.E. Am. J. Obstet. & Gynec. 99:808, 1967

Intrauterine devices: Physiological and clinical aspects. WHO technical report series No 379, 1968.

Lippes, J. and Kenmore, N.Y. Personal communication, 1970

Tamada, T., Okagaki, T., Maruyama, M. and Matsumoto, S. Am. J. Obstet. & Gynec. 98:811, 1967

Tietze, C. Studies in Family Planning 36:11, 1968. Population Council publication, New York.

Zipper, J.A., Tatum, H.J., Pastene, L., Medel, M. and Rivera,M. Am. J. Obstet. & Gynec. 105:1274, 1969.

STERILITY I. EXPERIMENTAL

GLYCOGEN IN SPERMATIDIC ARREST

A. Fabbrini, M. Re, M. Savioli and G. Spera

Istituto di Patologia Medica II, Università di Roma

Roma, Italy

Glycogen probably plays an important role in the maturation of germinal cells. Intratubular glycogen is abundant in the prepubertal stage; it diminishes remarkably at the beginning of puberty, reappears during the period of complete sexual maturation and the highest increase is found in the tubules in involution. Glycogen is abundant at some stages of spermatogenesis and almost completely absent in others. In our previous paper (1) we referred to a large accumulation of PAS positive material in the rat testis 7-10 days after a biopsy.

In the human testis considerable variations in glycogen content have been observed in pathological conditions. On the basis of these findings a study has been carried out on the behaviour of glycogen in spermatidic arrest.

MATERIAL AND METHODS

Testis biopsy was carried out on 128 patients on account of sterility or oligospermia, 33 of whom resulted affected by spermatidic arrest. 10 biopsies were performed on normal subjects for control studies. The specimens were fixed in Bouin's solution and paraffin slices were stained with PAS-haematoxylin. Control slices were pre-treated with α-amylase. The stages of spermatogenesis were determined according to Clermont (2). One tubule out of every three tubules was chosen at random. Sometimes we had difficulty in determining the stage on account of incomplete sequence of the cells, which may be observed in spermatidic arrest. The amount of glycogen was determined as arbitrary units: 0, +, ++, +++; 0 representing complete absence of histochemically detectable glycogen. From each specimen, 30 tubules were studied. The frequency of each spermatogenic stage and intensity of PAS reaction was determined.

RESULTS

The curve of the frequency of the spermatogenic stages in the spermatidic arrest does not greatly differ from normal. In both normal and pthological testes stages 2-4-6 are comparatively shorter than stages 1-3-5 (fig. 1); however these results differ from the findings of Clermont, who reported a longer duration of the 2nd compared to the 3rd stage.

The curves of glycogen in normal testis show the highest positivity at the 1st and 5th stages of spermatogenesis; on the contrary a plateau of PAS-positivity can be observed in the spermatidic arrest (fig. 2).

DISCUSSION

These findings give clear evidence of a close relationship between pathological spermatogenesis and the behaviour of glycogen. In a previous paper (3) we showed how the behaviour of gly-

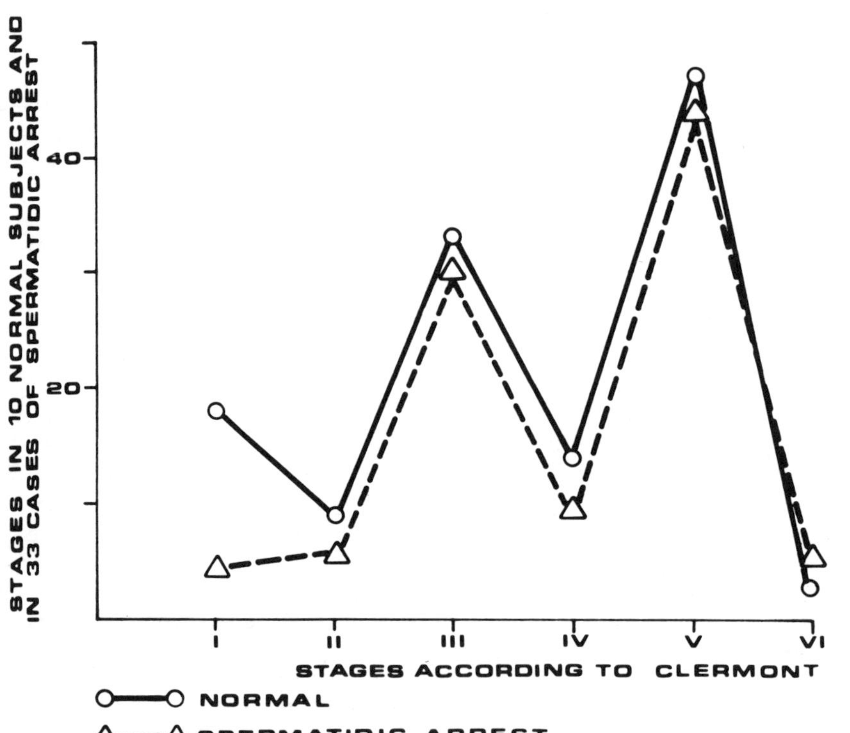

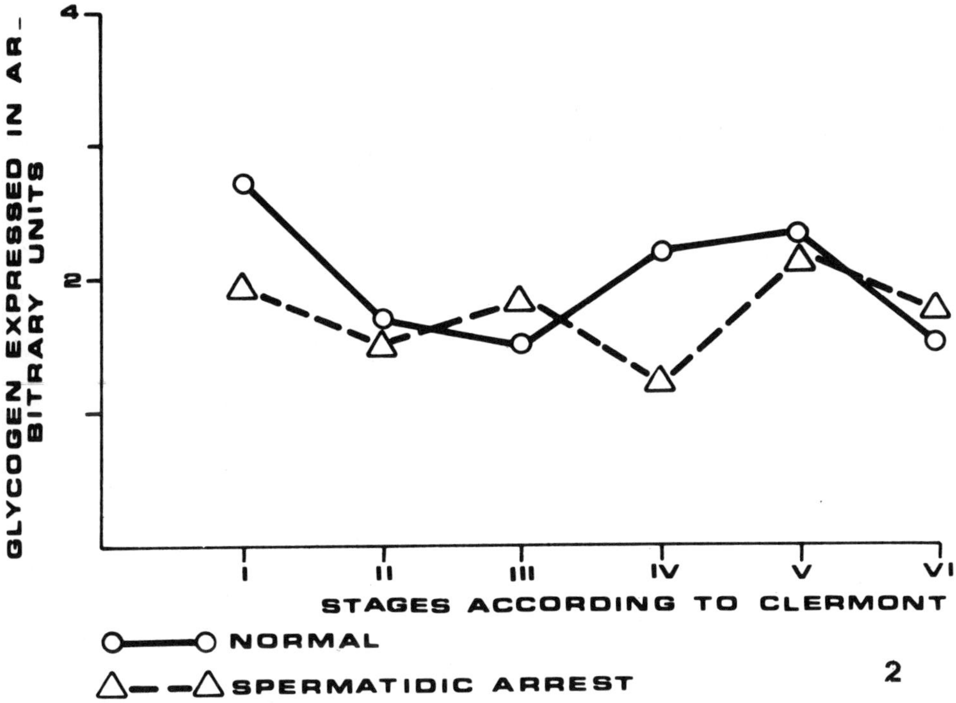

2

cogen was related to the development of spermatogenesis. The variation in intratubular glycogen would therefore appear to be directly related to the high energy required in nucleic acid synthesis in some phases of spermatogenesis. The fact that there is no fall in the amount of glycogen in the 3rd stage of spermatogenesis may mean that the energetic material is not used. Therefore, in spermatidic arrest there may be a defect in the biochemical maturation of the germinal cells even though the maturation curve of the germinal elements appears to be normal. This hypothesis seems to be supported by experimental data (4) which have revealed that infertility is often due to an arrest in the maturation of deoxyribonucleic acid in the spermatides on account of the defect in binding of the DNA-PO_4 groups, or an intrinsic biochemical anomaly in the nuclear protein with transformation of the histone into a protein which on account of its arginine content is a more basic protein.

Biochemical modifications are probably responsible therefore for the abnormal behaviour of the glycogen curve in spermatidic arrest following which there could be a blocking in the maturative mechanism resulting in an accumulation of glycogen especially in those stages in which there is an increased energy requirement.

REFERENCES

(1) Re,M. et coll. Folia endocr.18,662-667,1965
(2) Clermont,Y. Am.J.Anat. 112,35-51,1963
(3) Fabbrini,A. et coll. J.Endocr. 43,499-506,1969
(4) Gledhill,B.L. et coll. J.Reprod.Fert. 12,575,1966

DIHYDROTESTOSTERONE (5 α-ANDROSTAN-17-β-OL-3 ONE) IN THE EPIDIDYMIS OF RATS

Jan Th. M. Vreeburg and Jacob H. Aafjes

Department of Endocrinology, Growth and Reproduction. Rotterdam Medical Faculty, Rotterdam, The Netherlands

From a variety of studies it is becoming clear that androgenic influences at the cellular level are mediated by the conversion of testosterone to dihydrotestosterone and that dihydrotestosterone is possibly the major steroid necessary for androgenic phenomena in accessory sex organs.

To study the metabolism of testosterone (T) to dihydrotestosterone (DHT) in vivo we chose the epididymis as a

Table 1

Testosterone (T) and dihydro-testosterone (DHT) concentrations in epididymides of 3-month old rats.

No.	Weight (g)	T(ng/g)	DHT (ng/g)
12	0.82	lost	54.9
14	0.89	6.7	38.2
34	0.98	5.1	39.8
56	1.04	dirty	46.2
84	0.87	3.3	50.6
100	1.09	5.3	30.3
106	0.91	5.8	47.3
Mean	0.94 ± 0.04		44.0 ± 3.2
15	0.85	4.7	58.8 (4 heads)
16	0.83	<1	19.2 (4 tails)

target organ, since it was recently shown that T can be transformed to DHT in vitro in this organ (1).
The method we use for the estimation of T and DHT is a gas chromatographic method essentially the same as described by Van der Molen and Groen (2), which gave in our hands about the same precision as stated by these authors and appeared to be sensitive enough.

A major finding we obtained, was that the concentration of DHT was much higher in the epididymis then was T. This is shown in table 1, which gives these concentrations together with the weights of the left and right epididymis combined. It shows also that most of the DHT is contained in the head of this organ.

For comparison we studied the testis of the same animals where just the opposite results were obtained (table 2).

Table 2

Testosterone (T) and dihydro-testosterone (DHT) concentrations in testis of 3-month old rats.

No.	Weight (g)	T(ng/g)	DHT (ng/g)
11	1.10	156	7.3
13	1.25	166	4.8
33	1.64	91	13.4
35	1.64	108	10.4
83	1.35	63	14.1
Mean	1.40 ± 0.11	117 ± 20	10.0 ± 1.8

The concentrations of T were considerably higher than those of DHT.
We were excited to find from these results that the concentration of DHT was more than 4 times higher in the epididymis than in the testis. As mentioned already DHT is probably derived from T. We therefore were interested to see what would happen if the testes were stimulated to a higher T production.

To study this, intramuscular injections of 100 I.U. of human chorionic gonadotrophin (Pregnyl, Organon) were given to five rats, 3 month of age. After predetermined time intervals had elapsed these animals were sacrificed and used for analysis.

Fig 1 Concentrations of testosterone and dihydrotestosterone in the testis following HCG treatment.

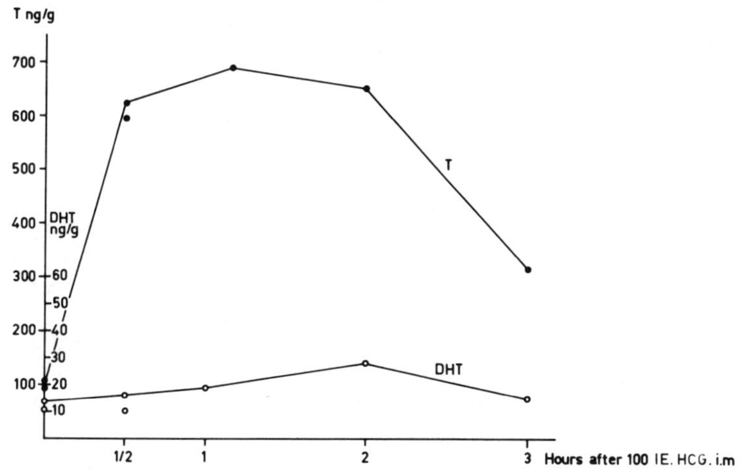

Fig 1 shows the effect of this treatment on the testis over a four hour time period. At each indicated time, the points are from a single animal. From the first two animals both testes were analysed and the results are given in the figure. From the other animals only one testis was used.

Within half an hour the T concentration rises considerably. The DHT concentration also rises although at a slower rate. Fig 2 shows what happened in the combined left and right epididymis from these same animals.

Fig 2 Concentrations of testosterone and dihydrotestosterone in the epididymides following HCG treatment.

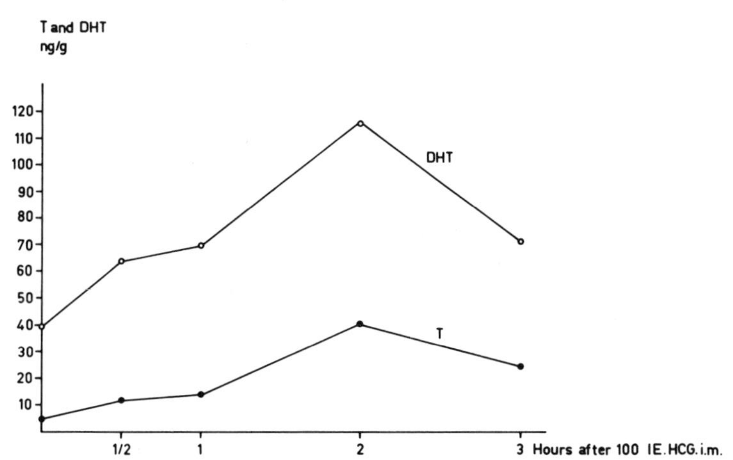

In this organ the increase in DHT is paralleled by an increase in T, which makes it probable that in fact it is the increase in T which causes the rise in DHT. If this is indeed the case the increase in concentration of T by the epididymis seems to be caused by a rather slow process compared with the fast increase of testosterone production in the testis as shown in figure 1.

In an additional experiment which lasted for 8 hours these results were confirmed. In the period from 1-8 hours after HCG administration the concentration of DHT in the epididymis was 93.2 ± 8.2 ng/g (mean of 6 animals ± s.e.m.) significantly higher ($p < 0.001$) then the concentration present in the epididymis of untreated animals (table 1). In the testes of these animals the DHT rose to 35.0 ± 6.91 ng/g which is also significantly higher ($0.025 > p > 0.01$) then the concentration in the unstimulated organ given in table 2.

Summarising we can state:
that the concentration of DHT in vivo is high in the epididymis, compared with the concentration of this compound in the testis and compared with preliminary results we obtained for the prostate and the seminal vesicle. We even do believe that this concentration might be the highest of the whole body.

And secondly that this concentration can even be increased when more T becomes available to this organ.

In connection with the high androgenic potency of DHT and the fact that maturation of spermatozoa takes place predominantly in the head part of the epididymis one could easily speculate that the presence of DHT might be of importance for this maturation. We hope that further studies will throw a little more light on this relatively badly understood organ.

REFERENCES

INANO, H., A. MACHINO & B.I. TAMAOKI, Endocrinology 84, 997 (1969).

MOLEN, H.J. v.d. & D. GROEN, J. Clin. Endocr. 25, 1625 (1965).

FSH-RELEASING ACTIVITY OF CLOMIPHENECITRAT

S. Heller and R. Blobel

Univ. Frauenklinik, Tübingen

Germany

In human female clomiphene induces ovulation but it inhibits fertility in rat. Clomiphene induces a rise of total gonadotrophic activity in urine and of FSH- and LH-activity in plasma in man. In contrast, an inhibition of pituitary gonadotrophin release probabely due to blocking of hypothalamic or pituitary centers, was discribed by Holtkamp (2). His findings were at variance with those of Igarashi et al. (3) who found in immature male rats a significant rise of both FSH and LH levels in the plasma. Stereotactic implantations of clomiphene crystals into the median eminence was followed by a decrease of pituitary FSH content. The following experiments shall demonstrate the mechanism.

We tested clomiphenecitrat of its FSH- and LH-activity by a modified method discribed by Ramirez and Mc Cann. Female Wistar rats of approximately 200 g body weight, kept under standard conditions, were bilaterally ovariectomized 4 weeks prior to the experiments. The pituitary of the animals were blocked with 25 mg progesteron and 50 µg estradiolbenzoat by subcutaneously injection 3 days prior to the assay. Clomiphenecitrat was injected in 0,5 ml physiological sodium chlorid solution and by another injection 400 E heparin in 0,5 ml physiological sodium chlorid solution into the tail vein. After 30 minutes the animals were anesthetized with ether and exsanguinated by punction of the vena jugularis. The plasma of 5 animals were pooled. Plasma FSH-activity was assayed by the method of Steelman and Pohley (5). For LH assay the animals were exsanguinated after 90 minutes. Plasma LH-activity was assayed by the ovarian ascorbic acid depletion by the method of Parlow (6).

Doses of clomiphenecitrat of 0,2 mg to 10 mg were assayed by rats with 200 g body weight. The effect of clomiphenecitrat upon the plasma FSH-activity is shown on figure 1.

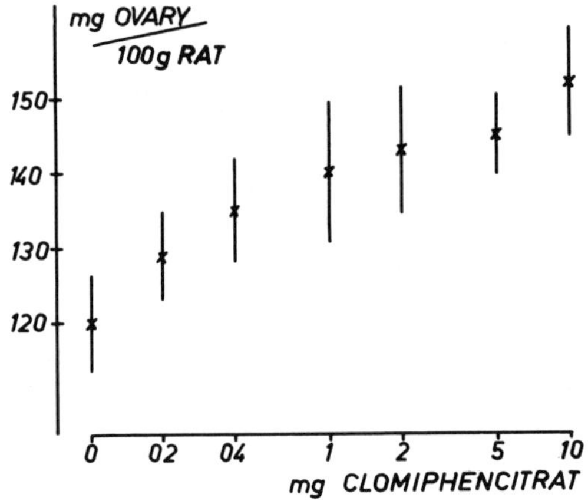

Fig. 1: Plasma FSH-activity of ovariectomized estrogen-progesteron blocked rats after injection of clomiphenecitrat.

Dosage increases of 0,2 mg each to a total of 10 mg resulted in a steady augmentation of ovarian weight and therefore of plasma FSH level. The differences observed between the groups injected with 0,2 mg and 2 mg of clomiphenecitrat were statistically significant (P = 0,01).

There was no effect of clomiphenecitrat upon the plasma LH-activity measured by ovarian ascorbic acid depletion test.

It is concluded therefore, that the increase plasma FSH-activity observed in ovariectomized, estrogen-progesteron blocked rats after i.v. injection of clomiphenecitrat was due to an effect of this compund on the releasing mechanism of FSH.
The results presented here are in agreement with the findings of Igarashi et al. (3). In contrast to the experiments to Baier (1) we couldn't find after an injection of 300 ug clomiphenecitrat and more a decrease of plasma FSH-activity.

Summary:

Plasma FSH-activity is due to the dosage level of clomiphenecitrat. But we couldn't find an increase of LH-activity in plasma due to the dosage level of clomiphenecitrat.

Literature:

1. Baier, H. and Taubert, H.D.
Endocrinology 84, 946 (1969)

2. Holtkamp, D.E., Greslin, J.G., Root, Ch.A. and Lerner, L.J. Proc. Soc. Exp. Biol. Med. 1o5, 197 (196o)

3. Igarashi, M., Ibuki, Y., Kubo, H., Kamioka, J., Yokota, N., Ebara, Y. and Matsumoto, S.
Amer. J. Obstet. Gynec. 97, 12o (1967)

4. Ramirez, V.D., Naller, R. and Mc Cann, S.M.
Endocrinology 73, 193 (1963)

5. Steelmann, S.L. and Pohley, F.M.
Endocrinology 53, 6o4 (1953)

6. Parlow, A.F.
in Albert, A.: Human Pituitary Gonadotropins
Springfield Charles C. Thomas (1961)

STERILITY II. CLINICAL

ESTIMATION OF THIOCYANATE LEVEL IN UTERUS

PRELIMINARY REPORT

ZDENEK KUBES, M.D.
ASSISTANT PROFESSOR, DEPT. OF GYNECOLOGY AND OBSTETRICS, NEW JERSEY COLLEGE OF MEDICINE, U.S.A.

Subtle aberrations in the biochemistry of the reproductive tract may be the cause of elusive infertility problems in the "otherwise normal couple." Intergrated hormonal influences regulate the cyclic metabolic changes necessary to provide the proper chemical environment for both the maintenance of sperm viability, and also the fertilization and subsequent development of the ovum.

There are many papers dealing with thiocyanate treatment of hypertension and with the significance of thiocyanate in the gastro-intestional tract. In our opinion the only paper dealing with thiocyanate levels in the uterus relates to a bacteriological study (1). Many papers deal with the fact that the sperm needs anaerobic conditions to fertilize the ovum (2,3,4,5,). No paper describes the relation between thiocyanate level in the uterus and infertility problem.

If anaerobiosis plays an important role, as has been suggested, then perhaps the appraisal of viability of spermatozoa should be made under anaerobic conditions similar to those that may be expected to be present in the female reproductive tract. This factor may be important in understanding the reasons for the infertility of the otherwise normal couple.

The cyclically changing endometrium and endocervix contain varying amounts of thiocyanate (1). It is known that varying amounts of thiocyanate destroys pathogenic bacteria in the gastrointestinal tract (6,7). If anaerobiosis plays an important role as had been suggested in problems of fertility, we may anticipate that the higher concentration of thiocyanates within the uterus and seminal fluid allows an optimum environment for fertilization (cervical permeability and sperm motility).

For the immediate clinical evaluation of therapy with Thiocyanate containing drugs and appropriate diet in those cases with a low pre-therapeutic Thiocyanate level we used the Sims-Huhner Test. Schreiber's method (7) modified for spectrophotometer by bowler was used in 32 patients (17 patients with two or more children) for determination of the Thiocyanate level in the uterus. The average Thiocyanate level in uterine cavity of infertile patients was 0.15 mg% and in fertile patients the level was 0.40 mg%. The therapy results will be published after comparative studies in greater number of patients will be available.

REFERENCES

1. Thies, K., Lange, R., Eigenschutz der weiblichen Geschletsteile und des Bauchfells. Arch. Gynaek. 172: 359-366, 1942.

2. Gross, Milton, Biochemical Changes in the Reproductive Cycle, Fertility and Sterility, Vo. 12, No 3,245, 1961.

3. MacLeod, J. The Role of Oxygen in the Metabolism and Motility of Human Spermatozoa. Am J. Physiol. 138: 512, 1943.

4. MacLeod, J. Sulfhydryl Groups in Relation to the Metabolism and Motility of Human Spermatoza. J. Gen. Physiol. 34:705, 1951.

5. Mann, T. Anaerobic Metabolism of Spermatozoa. Nature 156:80, 1945.

6. Lockemann, G. Ulrich W., Ueber die Bedutung des Rhodangehaltes des Magensaftes. Dtsch. med. Wschr. 1900, 1930.

7. Schreiber, H. Ueber den Rhodangehalt im menschlichen Blutserum. Biochem. Z. 163: 241-251, 1925.

LYMPHOCYTE TRANSFORMATION IN WIFE-HUSBAND MIXED LEUCOCYTE CULTURES IN NORMAL AND PATHOLOGICAL PREGNANCIES

I HALBRECHT AND L KOMLOSH

DEPARTMENT OF OBSTETRICS AND GYNECOLOGY, HASHARON HOSPITAL, PETAH TIQUA, AND THE RESEARCH LABORATORY OF HUMAN REPRODUCTION AND FETAL DEVELOPMENT, TEL AVIV UNIVERSITY MEDICAL SCHOOL

Because of the paternal genetic characteristics which the fetus possesses at least partly and which differ from those of the mothers´ the fetus is to be considered as a homograft to the mother. Such a genetic disparity existing between mother and the fetus would lead to the rejection of any other homograft implanted in the uterus or in any other of the maternal organs.

Why is it, therefore, that the fetus escapes normally immunological destruction? On the other hand we may also ask ourselves whether disturbed pregnancies and particularly abortion and unexplained intrauterine fetal death at any stage of the gestatation may not be due to a disruption of the natural immunological defense mechanism which the mother develops during a normal pregnancy and which is aimed to protect the fetus from destruction and rejection.

Medawar and other investigators (1) have extensively reviewed the problems associated with the immunology of pregnancy and have suggested the following explanations which might account for the survival of the natural homograft in the uterus, the conceptus.

1.- The fetus is anatomically completely separated from the mother.
2.- The fetus is antigenetically immature.
3.- The mother is immunologically inert during pregnancy.
4.- The uterus is an "immunologically privileged" organ.

All these explanations have not been substantiated by the investigations of most of the other researchers.

More recently Kirby (2) has suggested that after implantation the conceptus develops its protective barrier in as much as the trophoblast layer which intervenes between the embryo and the mother becomes invested with a pericellular coat ot SIALOMUCIN which seems to mask the transplantation antigens of these cells so the mother does not recognize them as foreign and does not react against them.

Simon and Russel (3) have postulated that the implanted egg is also protected from the immune attack by a non-cellular barrier, the zona pellucida which according to investigations done by Soupart and Moyes (4) contains sialic acid.

Curie and Bagshave (5) consider that it is the sialic acid which is responsible for the immunological properties of the fibrinoid coat by exerting a strong electro-negative effect which repels the negatively charged lymphocytes.

It is admitted that there are two populations of lymphoid cells that make up the body's immune system. Both populations arise from the same stem cell but are differentiated along different pathways at different sites and occupy different regions of the peripheral lymphoid tissue.

The so-called thymus-dependant cells of small lymphocytes which after leaving the thymus circulate through the blood and the lymphatic system are responsible for the so-called cellular immune response. They are involved in graft rejection, delayed hypersensitivity and graftversus-host reactions.

The thymus-independent cells gain access to the circulation as do the thymus-dependent cells and finally come to lodge in areas of the peripheral lymphoid system separated from those occupied by the thymus-dependant cells which are seeded in the deep cortical regions of the lymph nodes and in the spleen. It is not yet known where they undergo differentiation but they are responsible for the immunglobulins and the production of immunantibodies IgM, IgC and IgA as well as more specific antibodies.

The question to be asked is how these two, the cellular and humoral, immune systems of the body behave during gestation and how are their activities cancelled in order to prevent damage to the conceptus and if damage is produced which of the two systems may be responsible for it.

The answer to these questions may give us the clue to a better understanding of many pathological conditions of pregnancy and perhaps also of malignancy in general and of malignancy of the trophoblast cell in particular.

Feto-maternal ABO or Rh blood group incompatibility is known to damage the red blood corpuscles of the fetus causing hemolysis and icterus.

Since the Rh antigens have so far been found only in the erythrocytes no other damage in the fetus ought to be expected in case of feto-maternal Rh blood group incompatibility.

In cases of sensitisation of mothers against the A or B blood group antigens of their fetuses on the other hand damage to any other fetal tissue might be expected, since these antigens are ubiquitous in all fetal tissues and organs.

We have indeed been able to show (6) that iso-immunantibodies anti-A and anti-B exert cytotoxic effects on placental cells in tissue cultures.

Laman and Harod (7) have on the other hand demonstrated that maternal cytotoxic antibodies reach the fetus but do not damage it. Maternal antibodies do not therefore necessarily have an adverse effect on the fetus in vivo.

It has furthermore been shown that twenty per cent of mothers who have had at least two pregnancies have leuco-agglutinins in their circulation, yet these antibodies do not seem to induce changes in white cell counts or other immunological reactions in the infant normally. But whether they are always without ill-effect on the fetus remains to be demonstrated by further extensive investigations.

Whether incompatibility of transplantation antigens between mother and fetus elicits in the mother only humoral immune antibody reactions as does feto-maternal blood group incompatibility or whether cellular, for the fetus potentially more damaging immunologic reactions are also involved, has so far not been clarified.

In our studies we have used lymphocyte transformation in wife-husband mixed lymphocyte cultures to detect the existence of histo-incompatibility between mother and fetus who, as already mentioned, possesses paternal characteristics and in our previous communication (8) we have already reported on our results of this test in abortions and hydatidiform moles.

In our present communication we wish to report on further results which we have obtained by using this test in various pathological conditions of gestation.

Material and methods:

The percentage of transformed cells was determined in wife-husband mixed lymphocyte cultures in 165 cases of normal and abnormal pregnancies. Lymphocyte transformation was at the same time also examined in cultures of the wife only. (See fig 1).

In 25 out of the 165 cases the pregnancy was normal and in 140 cases we dealt with various abnormal conditions of gestation.

Fourteen normal fertile couples served as controls.

The hundred and seventy-nine couples which were the objects of our studies were divided in the ten following groups:

Group one included 14 women who have had 2-4 normal pregnancies and their husbands.

Group two included 25 women who were 1-3 days after a normal delivery and their husbands.

Group three included 14 cases of premature deliveries.

Group four included 20 cases of threatened abortions.

Group five included 32 cases of premature deliveries.

Groups six and seven included 32 women after their second abortion and 11 women who had miscarried 3-5 times respectively.

Group eight concerned 12 cases of hydatidiform moles.

Groups nine and ten finally included 16 women who were delivered of macerated fetuses and 15 patients who were suffering from toxemia of pregnancy respectively.

RESULTS:

Table 1

NO. OF CASES:	SPECIFICATION OF GROUPS:	PERCENTAGE OF TRANSFORMED CELLS: MIXED CULTURES:		WIFE CULT ONLY:	
		AVERAGE:	RANGES:	AVERAGE:	RANGES:
14	CONTROL	6,7	2,6- 9,6	6,9	4,3- 9,3
25	NORMAL DELIVERIES (PUERPERIUM)	6,2	2,4-13,8	4,4	0,4-10,6
14	PREMATURE DELIVERIES "	13,8	6,4-37,0	5,1	0,9- 8,7
20	THREATENED ABORTIONS	13,9	2,7-33,0	7,8	2,6-10,0
32	AFTER FIRST SPONT ABORTION	15,7	4,2-33,9	6,7	1,6-11,0
20	AFTER SECOND ABORTION	19,3	2,8-39,0	6,1	1,7-10,4
11	AFTER 3-5 ABORTIONS	20,2	10,0-33,0	7,1	3,6- 9,0
12	HYDATIDIFORM MOLES	16,7	6,0-35,0	6,3	1,7-11,0
16	AFTER DELIV. MACERATED FETUS	16,6	6,0-26,0	9,0	2,0-10,0
15	TOXEMIA OF PREGNANCY	9,1	2,5-16,1	5,3	1,0-10,8

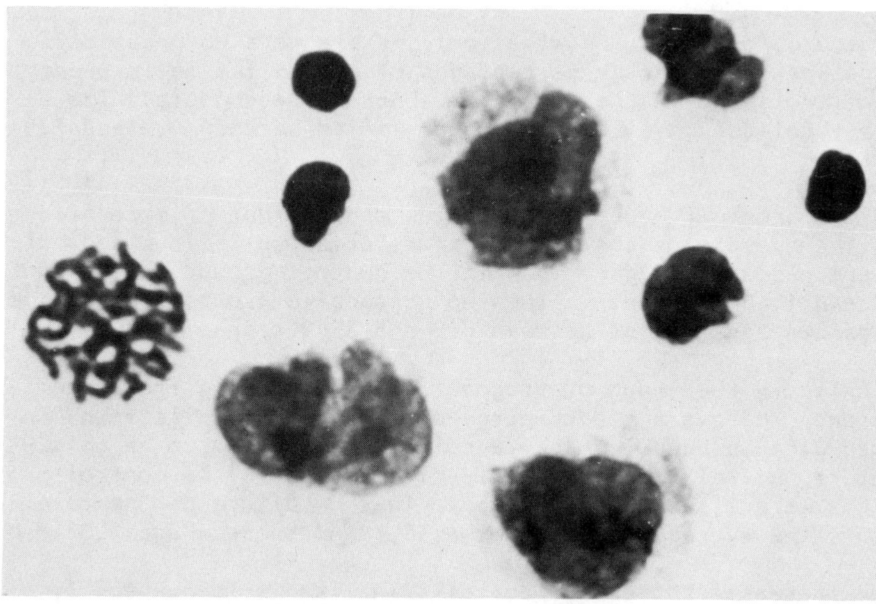

Fig 1. Transformation of lymphocytes into blast-like cells in a mixed wife-husband leukocyte culture in habitual abortion.

The percentage of transformed cells in mixed wife-husband lymphocyte cultures was found to be almost alike in the control group and in the group of women after normal deliveries. The percentages were on the average 6,7 and 6,2 respectively for both groups. The ranges were 4,4-9,0 and 0,4-10,6 respectively.

In the five groups of disturbed pregnancies, namely: premature deliveries, threatened abortions, and one, two or more complete abortions the averages of percentage of the transformed cells were much higher than in the normal groups, They were 13,7, 13,9, 15,7, 19,3 and 20,2 in the five respective groups. So were also the ranges in these groups. They varied between 2,7 and 39,0.

In the cultures of the wives only, there were no great differences between the average percentages of blast-like cells present, in the controls and in the groups of abnormal pregnancies. The average percentages in all these groups varied between 4,4 and 7,9 per cent.

In the group of hydatidiform moles and in that of macerated fetuses the average percentages of transformed cells in mixed cultures were also much higher than in the control group. They amounted to 16,7 and 16,6 per cent in the two respective groups. The ranges of the percentages varied between 6,0 and 35,0 respectively.

Finally in the group of pregnant women suffering from toxemia of pregnancy the average percentage of transformed cells found in the mixed wife-husband lymphocyte cultures was found to be only 9,1 per cent, i. e. relatively slightly higher than in the control groups. The ranges were 2,5 to 16,1 per cent. In the culture of the toxemic women only the average percentage was 5,3 and the average 1,0-10,8.

The increase of transformed cells in mixed wife-husband lymphocyte cultures in the groups of disturbed pregnancies as compared with that found in normal pregnancies or non pregnant women and their husbands was even more conspicuous in table 2 in which the classification was made according to whether more or less than 10 per cent of transformed cells were found in the cultures.

In none of the mixed wife-husband lymphocyte cultures of healthy fertile couples with normal pregnancies did we find more than 10 per cent transformed cells whereas in cases of 3-5 abortions in the past of the wife none of the cultures showed percentages of less than 10 per cent of blast like cells. In all other groups of cases with pathological pregnancies such as premature deliveries, threathened abortions, complete abortions, hydatidiform moles and macerated fetuses the average percentage of transformed cells in the mixed cultures above 10 was from two and a half to five times more frequent in abnormal pregnancy groups than in the control group.

Table 2

NO. OF CASES:	SPECIFICATION OF GROUPS:	PERCENTAGE OF TRANSFORMED CELLS: ABOVE 10 %	BELOW 10 %
14	CONTROL	NONE	14
25	NORMAL DELIVERIES (PUERPERIUM)	5	20
14	PREMATURE DELIVERIES "	10	4
20	THREATENED ABORTIONS	15	5
32	AFTER FIRST SPONT ABORTION	25	7
20	AFTER TWO SPONT ABORTIONS	15	2
11	AFTER 3-5 SPONT ABORTIONS	11	NONE
12	HYDATIDIFORM MOLES	10	2
16	AFTER DELIVERIES MACERATED FETUS	12	4
15	TOXEMIA OF PREGNANCY	6	6

Discussion:

Our studies were based on the assumption that although nature has put up a special and complex defense mechanism to protect the mammalian fetus against any immunological attack coming from its host, the mother, there is still a possibility that histo-incompatibility may play an important role in the pathogenesis of disturbed pregnancies and especially in cases of rejection of the conceptus in the various stages of gestation.

We therefore endeavored to detect the existence of a histo-incompatibility between wife and husband and consequently between the mother and her fetus by using mixed wife-husband lymphocyte cultures for the determination of the percentage of transformed blast-like cells in these cultures which is generally considered as being a fairly reliable test for the determination of histo-compatibility between recipient and donor of transplants.

The results which we have obtained seem to give evidence of an at least much higher incidence of histo-incompatibility existing between wife and husband in cases of disturbed than in normal pregnancies.

At this stage of our investigations we are not entitled to reach more far reaching conclusions as to the pathogenic role of feto-maternal histo-incompatibility in rejecting or damaging the fetus despite the manifold evidences of immunologic reactions of the mother against the blood group as well as the transplantation antigens of her fetus. Nor are we in the position to tell which of the two immunological mechanisms the humoral or the cellular, intervene in the eventual process of rejection of the conceptus if at all.

More extensive investigations will be needed to solve the problem of feto-maternal histo-incompatibility and its role in disturbed pregnancies.

Summary and conclusions:

Mixed wife-husband lymphocyte cultures were done in 25 normal and 140 cases of abnormal or pathological pregnancies and the percentage of transformed cells was determined. The cultures of 14 cases of normal fertile couples without pregnancies served as controls.

The average percentage of transformed cells found in cases of premature deliveries, threatened abortion, complete abortion, hydatidiform moles, and macerated fetuses was found to be considerably increased compared with that found in normal pregnancies or control cases.

In view of these results it is suggested that histo-incompatibility between mother and fetus may play an important role in the pathogenesis of rejection of or damage to the conceptus.

REFERENCES

1. Medawar, P.B., :Some immunological and endocrinological problems raised by evolution of vivparity. Sympos. Soc. Exp. Biol. 7:320-30 (1953).

2. Kirby, D.R.S., Billington, W.D., Bradbury, S., and Goldstein, D.J.: Antigen barrier of the mouse placenta. Nature 204:548-549 (1964).

3. Simmons, R.L. and Russell, P.S.: Immunological problems of Pregnancy. Am. J. Obst. & Gyn. 85:583-593 (1963).

4. Lamman, S.T. and Harod, L.: Homograft Immunity in Pregnancy. J. Exp. Med. 122:572-76. (1965).

5. Soupart, P. and Noyes, R.W.: Sialic Acid as a component of the zona pellucida of the Mammalian Ovum. J. Reprod. Fertil. 8:251-253 (1964).

6. Bagshave, K.D. and Currie, G.A.: Immunogenicity of L 1210 Mouse Leukaemia cells after Treatment with Neuraminidase. Nature 218:1254-1255 (1968).

7. Halbrecht, I., Komlosh, L. and Gertner, C.: L'Action des Immuno-Anticorps anti -A et anti -B sur les Cellules Placentaires en Culture: Les Avortements du Premier Trimestre, Masson pp. 193-205 (1967).

UTERO-ABDOMINAL REFLUX AND FEMALE STERILITY: A CLINICAL STUDY

P.EHRLER, M.D.

Viktoriaspital CH - 3000 BERNE

The scientific papers of the last years concerning the removal of tubal obliterations in female sterility give mostly advices in surgical technique with or without additional medical therapy.However,they have to be considered as a one-sided therapy and they do not touch basic problems.With the aid of adhesiolysis, salpingolysis,salpingostomy and different methods forming a new calix ampullae,there can certainly be reached a series of positive results.On the contrary,some failures have to be noted,despite of the afore-mentioned methods.Re-occlusion was not a usual,but inevitable and a to often occuring complication.

Certain progresses have been made with hydrotubation,especially with the long-time hydrotubation of Fickentscher and Semm instilled a mixture of streptomycine/hydrocortisone/chymotrypsine. The application of a portio-pessary for hours inhibits the outflux of the instilled fluid; in addition to that the pressure in the direction of the tubal ends can be augmented,enabling the lysis of any obliterations.Swolin used very high doses,i.e.1000 to 2000 mg of hydrocortisone at a time when the internists still argued whether 5 to 10 mg could be tolerated by the organism.Both methods,as good as they might be,still include a certain quota of failure.

The question poses as before,in how this tendency of occlusion of the tubal ends could be overcome.What we know is the fact that tubal obliteration is basically caused by inflammatory processes spread of an intraabdominal inflammatory lesion that is appendicitis or lymphadenitis(endometriosis in this connection ought to be excluded).We know obliteration as a cause of hematogenous spread,tuberculosis being the classical example.We know obliteration on the basis of gonococcal infections,the example of ascending infection.Therewith by far not all causes of obliteration have been mentioned.The fact that the tubes in contrary to other hollow organs show after their operative correction such an extreme tendency to re-occlusion even a long time after the aforementioned primary infections has herewith not yet been explained; at a time when the latter can by no means be considered as damaging agents.Undoubtedly there is another factor for tubal occlusion.

By a simple method,but considering the matter from a new point of view,that factor could be traced by the aid of simple inspection and topographical anatomy,which led the operative therapy in a new direction.

What do we see and what do we establish regularly?We diagnose occluded tubes adhering with a significant regularity at the dorsal wall of the uterus and mostly lying deeply in the space of Douglas.Looking for an explanation for this caracteristic position and occlusion of the tubes,we aim at a typical phenomenon of the human organism with its hollow organs.Gastroenterologists know it in connection with hiatal hernia and the pediaters in connection with pyelonephritis.We mean the reflux-phenomenon,the reflux-disease.The presumption lies near that also in these hanging tubes with their tendency of occlusion a reflux could play a role,the utero-abdominal reflux of menstrual blood with more or less infected secretion of the uterus.An inquiry amongst gynaecologists and non Gynaecologists resulted as follows:
The non gynaecologists say: does reflux exist at all?
The gynaecologists agree to this by they have to be ready to take responsibility for the lack of knowledge of non-gynaecologists. For there was rarely talked and still less written about the problem.In anglo-american literature there is no reference to it.Undoubtedly we have to talk about it and to answer the questions:

a) Is there utero-abdominal reflux at all?
b) Has it a pathogenetic importance in formation of tubal occlusion and other pelveo-peritoneal changes?

ad a)

1 The acute abdomen with tenderness and leucocytosis. Laparotomy shows fluid blood in big quantities,as in extrauterine gravidity.Looking for the origin of bleeding there is not any except the tubes which have secreted this blood during menstruation.

2 If laparoscopy is being carried out during menstruation,the Douglas-secretion,which if found plentiful, is of redish colour.Laparoscopies during the interval show no longer red but yellow fluid(M.Berger,Berne). The erythocytes of the menstrual blood are hemolised within a short time.

3 Ever since one dissuaded from operations during menstruation.Nevertheless,if the patient is to be operated on,blood can in an unexpected number of cases be massaged retrogradely from the tubes into the abdomi-

nal cavity.Even spontaneous reflexes can be observed.

4 Prof.Closs,university of Bergen,observed by means of biochemical and electrophoretic investigations of the Douglas-fluid that it shows neither signs of inflammation(exsudation)nor signs of congestion(transsudation) but signs of blood,the menstrual blood.

ad b) The utero-abdominal reflux does not only exist,it has also pathogenetic qualities i.e.it is responsible for tubal occlusions and pelveoperitoneal lesions with the multiple symptoms resulting from the following facts:

1 It has always been stated that recidiving adhesions of the small bowles can only be cured after curing the lesions of the small pelvis especially after the uteral extirpation with its complete peripheral occlusion.

2 The incidence of pelvic inflammation in men is rarely seen, in woman it occurs very often.This cannot be brought in connection with tubal inflammation.

3 Veterinary medicine tells us that pelveoperitonitis as we see it in women does not exist in animals.In certain cases comparative anatomy and pathology relates this to the upright gait of human beings.

4 In carnivores and herbivores the tubes and ovaries are not situated in the Douglas,which is always free from infections,but in the abdomen. Vide illustration 1.

5 We are indebted to Dr.Dapunt(university clinic,dept.of gynaecology,Innsbruck)for providing the undermentioned casuistic case:

A 12 years old girl is being operated for acute appendicitis.A uterus duplex has been diagnosed.The one cornu is opened distally,the other one closed i.e.rudimentally developed. 2 years later the girl re-enters the clinic as an emergency case with acute abdomen and a tumor adjacent to the uterus.Laparotomy showed hematometra and hematosalpinx.The girl had in the meantime her first menstruation.The end of the tube near the closed cornu which at the time of appendectomy was intact and open,showed itself closed during second operation by the congested menstrual blood and its reflux. Vide illustration 3.

6 Literature according to our findings shows as a picture of the surgical gynaecological technique of Käser/Iklé: uteral malformations with distally closed cornu can result in bilateral and peripheral closed tubes as an effect of reflux of the menstrual blood. Vide illustrations 2 and 3.

There should be no doubt about the existence of the utero-abdominal reflux.However,we have misunderstood and underestimated it in its meaning if not overlooked completely.The best operative techniques are of no use in case the tubes are still exposed to pathological reflux.

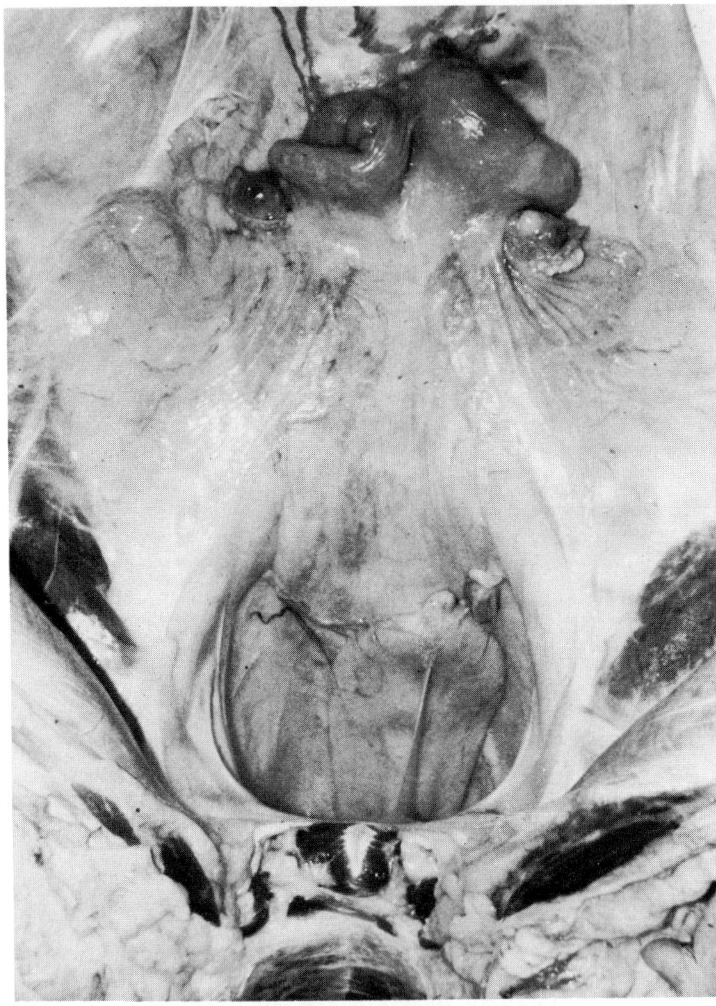

b

a

Illustration 1 : In the bovine species is

a) Douglas free from adhesions and
b) tubes and ovaries outside of Douglas, that is in the abdomen.

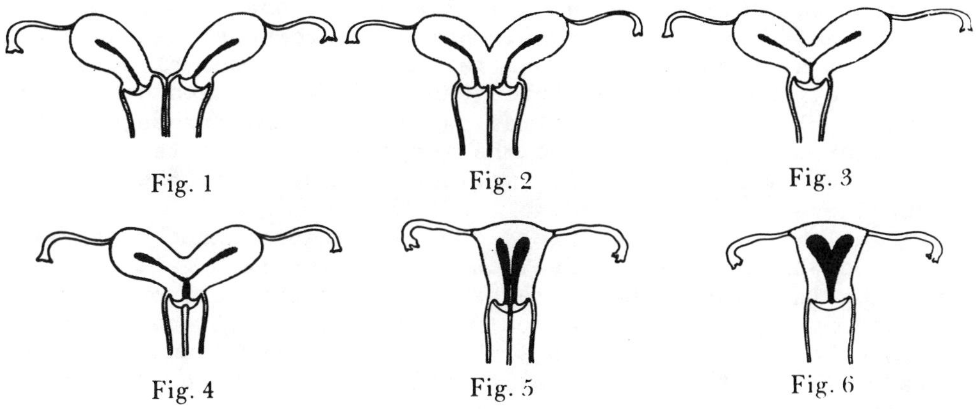

Illustration 2 and

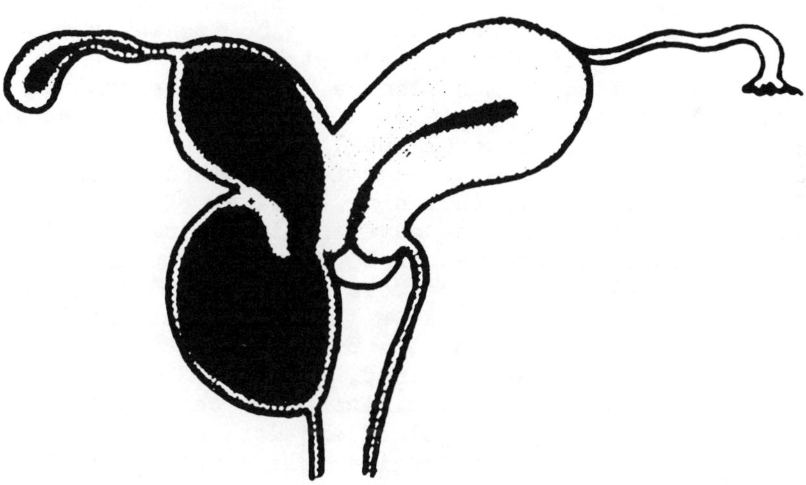

Illustration 3 :

Reflux of congested menstrual blood can also close the tube.

If these indications should not be sufficent,we may remember the french slogan: "Dans le Douglas pas de salut". The situation cannot be described more clearly and more comprehensive.The first causes of tubal obliterations as well as adhesions in the small pelvis are not only tuberculosis,gonorrhoea,appendicitis and lymphadenitis but also utero-abdominal reflux of menstrual blood and uteral secretions.Obstructed tubes can so be expression and cardinal symptom of the utero-abdominal reflux-disease,expression of the intramural insufficiency of the insufficient muscular portion of the utero-tubal region.This hypothesis leads us to the following questions:
Is this reflux of menstrual blood due to mechanical and topographical factors which are in connection with the upright gait of the human being.In animals where uterus and tubes are horizontally situated and in addition to that outside the Douglas,it cannot be stated.Should we not therefore try to elevate uterus and tubes out of this small pelvis(Douglas) a location well-known for the accumulation of infectious material?

Brunner suggested an operative procedure of elevatio uteri extraperitonaealis et pexia lateralis adnexorum.In order to prevent reflux and adhesion,the ligamenta rotunda are being elevated medially at their lateral origin and attached to the oblique fascia according to the method of Gillom/Simpso.After elevation the tubes take the direction verically to the room of Douglas,a condition which can be arrived at by attaching the mesosalpinx near the originating lig.rotunda.The procedure has been named Pinguin-Method due to the fact that the tubes now lying horizontally resemble indeed a standing pinguin with spread-out arms.

A further indication that the reflux of menstrual blood and uteral secretions has a certain pathological meaning results from sterility cases with peripheral tubal occlusion which have been operated according to the afore-mentioned method.A Norwegian-Swiss study(Brunner,Ehrler,Flury)including 38 cases of peripheral tubal occlusion with intact fimbriae gave the following results:
13 gravities could be obtained,in 10 cases the tubal ends were radiologically intact after 4 and more months,in 15 cases new obliteration of the tubes occured.With the aid of salpingolysis,salpingostomy and the pinguin-method the procedure of Swolin(hydrocortisone 1500 mg)has also been combined.

With an extremely high significance we can say that the utero-abdominal reflux of menstrual blood and more or less of uteral secretion is responsible for a considerable number of tubal occlusions. Although,this phenomenon,hardly ever mentioned,does in fact exist. The treatment of the refluxes by the pinguin-method seems to eliminate the pathogenetical fact of these refluxes.Furthermore,the formerly observed excessive tendency to re-occlusion of the ob - structed fimbries is being favourably influenced.

THE EXPERIMENTAL ARTIFICIAL OVIDUCT

Masanori Kirimura, Jozi Ono, and Motoyuki Hayashi

Dpt. Obst. Gyn., Toho University, 6-11-1, Omori, Ota-Ku
Tokyo, Japan

There are many ways of treating patients suffering from sterility caused by tubal damage. We have tried the following methods.

1. Insufflation
2. Hydrotubation
3. Medicamentous hydrotubation
4. Salpingoplasty

In case of total occlusion of the tubes, five procedures for establishing pregnancy have been tried.

1. Implantation of the ovary into the uterine cornua.
2. Transplantation of the other intact tube into the uterine cornua.
3. Formation of tubes using neighbouring tissues.
4. Artificial insertion of fertilized ova into the uterine cavity.
5. Application of an artificial oviduct.

The last method constitutes a problem which we want to present.

We made an artificial oviduct out of latex in order to produce a communication between tube and ovary. This prothesis consists of a spermiduct, a capsule around the ovary and the oviduct.

Since the artificial oviduct has no active function as an ovum pick-up and transfer of fertilized ova we considered the following method. After ovulation some cultural fluid was poured from the spermiduct with delicate pressure to help the movement of fertilized ova towards the uterine cavity.

Since it was not possible to observe the phenomenon of fertilization and cleavage inside of the peritoneal cavity, we did this in a chamber similar to the capsule of the artificial oviduct made out of latex. Cleavage was observed in 30-50 % of the ova.

A hysterosalpingogram was made two weeks after application, showing no adhesions in the tubal and capsular cavities and the passage of contrast medium into the peritoneal cavity.

Histological examination revealed that the ovary had become atrophied and that ovulation had ceased owing to the thickness of the ovarian cortex following two to four weeks of application. We, therefore, had to improve the material and the operative technique.

Supposing that the ovary might have some characteristic of reacting against being closely covered by the capsule we considered different ways of preventing this reaction, for instance through making pin-holes or openings in the capsule. We were able to find a new material, which was less irritating, softer and permeable. Its chief component is dimethyl polysiloxane and is generally called medical silicon rubber. Histological findings after using this material reveal only slightly fibrous proliferation of connective tissue of the cortex and a lot of normal follicles still exist.

It, therefore, is our opinion that an artificial oviduct made out of silicon rubber might offer a possibility to treat sterility caused by complete tubal occlusion.

THE EFFECTS OF MESTEROLONE ON PATIENTS WITH IMPAIRED SPERMATOGENESIS AND SEXUAL DISORDERS

Avraham Mor M D and Nathan Amit B Sc

Semenology Institute, Tel Aviv, Israel

INTRODUCTION

Recent studies indicate that Mesterolone (1 Methyl-5 androstane-17 -ol-3-one) is a valuable drug in the treatment of impaired spermatogenesis and some sexual disorders[1,2,3]. It now appears that Mesterolone may be more effective than many other frequently used oral testosterone preparations[4,5]. This steroid exhibits adequate androgenic and spermatogenetic effects with no apparent impairment of liver functions. In this communication we are reporting our results obtained on patients treated with Mesterolone, for various sexual and fertility disorders[6]. Previous studies on the efficacity of Mesterolone, were carried out mainly on experimental animals and on small number of patients[4,7,8]. To the best of our knowledge, we are the first to report on the effects of Mesterolone on a large series of patients, who were carefully evaluated as to their response to this steroid, from different clinical, hormonal and biochemical parameters.

PATIENTS AND METHODS

96 ambulatory patients, 21 to 50 years of age, who were seen at the Semenology clinic in Tel-Aviv, were selected on the basis of complaints ranging from sterility to impotence and lack of libido. Special care was exercised to include into the study patients free of any other recognizable endocrine disorders and who have not been treated previously with any other hormone. Detailed medical histories and clinical examinations were performed on all patients. In addition, the following tests were done before and after treatment: 1. Semen analysis consisting of measurements of volume of ejaculate, the number of spermatozoa, motility and morphology

of spermatozoa and presence of absence of spermatogenetic cells. 2. Fructose determinations in the ejaculate. 3. Liver function tests consisting to SGOT and SGPT assays.

Total gonadotropins, 17 keto steroids and 17 hydroxy steroids, were tested on 56 patients. Testicular biopsies were performed on patients. Potency, libido as well as the mental and physical condition of the patients were evaluated individually before and following treatment. The daily dose of Mesterolone (Proviron Schering A G Berline), administrated to all patients was 30 mg per day for a duration of 45 to 60 days.

The patients were divided according to their objective semen findings, as recommended at the second International Congress on Fertility and Sterility (Naples, 1956)[9].

Table No 1

EFFECTS OF MESTEROLONE (30 mg/day) ON 88 PATIENTS WITH IMPAIRED SPERMATOGENESIS

Clinical and Laboratory Diagnosis	Patients	Results					
		No of cases				No of Preg	% improved
		N $\uparrow$	$\uparrow$	=	$\downarrow$		
Aspermia and Azoospermia	36	1	10	24	1	9	30.5
Oligospermia and Hypozoospermia	43	17	13	9	4		69.7
Normospermia	5	5^{x}					
Hyperspermia	4	4^{xx}				1	100.0
Total	88	27	23	33	5	10	

N $\uparrow$ Improved to normal values

$\uparrow$ Improved to lower than normal values

= No change

$\downarrow$ Decrease in density, motility and morphology

x Remained Normal

xx Returned to Normal

Table No 2

SEMEN VOLUME AND FRUCTOSE LEVELS AFTER TREATMENT WITH MESTEROLONE

Clinical and Laboratory Diagnosis	VOL of EJAC (No of cases			% cases with Vol Incr	FRUCTOSE No of cases			% cases with Fruc Incr
	↑	=	↓		↑	=	↓	
Aspermia and Azoospermia	26	4	6	72	12	20	4	33
Oligospermia and Hypozoospermia	36	1	6	84	15	15	13	34
Normospermia	5	0	0	100	5	0	0	100
Hyperspermia	3	0	1	75	1	1	2	25
Pure Impotence	6	1	1	75	1	7	0	12.5
Total	76	6	14		34	43	19	

↑ Increase

↓ Decrease

= No change

RESULTS

Eighty eight patients with varying degrees of spermatogenetic deficiencies received daily 30 mg of Mesterolone, for a minimum period of 45 days. Of these, 22 patients returned to normal sperm following the treatment schedule, 23 patients showed spermogram improvements (Table 1), 5 patients with normal sperm values, remained normal after Mesterolone administration. The 4 hyperspermetic patients however, uniformely presented decreases in sperm densities with godd motility. Only 5 patients reacted with decreases in sperm density as compared to the pre treatment spermiograms (Table 1). The semen volume and fructose levels showed significent rises in all patient categories (see Table 2).

There were no changes in gonadotropin values, the 17 Keto Steroids however, rose slightly following treatment. 7 patients on which testicular biopsis were performed showed substantial rises in the number of Leydig cells. Liver function tests, SGPT and SGOT levels remained within normal limits, although slight temporary rises in the transaminase values were occasionally noticed. None of the patients in this series presented any side effects to this hormone. In 30 patients in this series who also complained of potency and/or libido disorders improved in 73 % and 60 % respectively. In fact, 18 patients from this group reported return to normal libido after treatment.

DISCUSSION

Most of the oral testosterone derivatives, available prior to the synthesis of Mesterolone, were by in large less than satisfactory in their effect on improving spermatogenesis[5]. The results outlined above, show that many of the deficiencies of other 17 methyl testosterones were overcome by Mesterolone. We found that Mesterolone treatment improved spermatogenesis in 52 % of patients with impaired or absent sperm densities, 9 of these, resulted in pregnancy. It is notable that there was no decrease in sperm counts of normal patients, while all hyperspermic patients returned to normal. In addition to these overall superior results, we did not observe rebound phenomena in any of our patients. The slight increase of 17 keto steroids following Mesterolone treatment suggests, in the absence of a better hypothesis, that this may be due to augmented testosterone activity.

The increase in the numbers of Leydig cells found in post treatment biopsys [10,11, 12, 13] specimens, is another indication of the androgenic effect of Mesterolone. The gonadotropin levels, which remained stable throughout the treatment period [14, 15] offer a good explanation for the lack of occurence of rebound phenomena.

These effects, coupled with the absence of both hepato-toxicity and of side effects certainly point to the therapeutic usefulness and potential of Mesterolone.

SUMMARY

96 patients with different spermiograms and variable potency and libido disorders, were treated with Mesterolone (Proviron). The effects of treatment showed good to moderate results in most of the parameters of semen analysis studied.

REFERENCES

1. Schellen T M C M, Results with Mesterolone in the Treatment of Disturbances in Spermatogenesis. Andrology 2, 1 1970.

2. Niermann H and Noltings. Oral Methylandrostane in Fertility Disorders. Hautarzt 17, 8:358 (1966).

3. Schirren C. Fertilitatsstorungen des Mannes, Beitr Sexualitatsforsch, H 22, F Enke, Stuttgart 1961.

4. Bibliography on Oral Administration of Androgens. Bibliography of Reproduction, Vol 14 No 4 (Apr 1969).

5. Parkes A S. Proc 8 Int conf. IPPF Santiago 67, 502, 1967. (IPPF, London).

6. Hornstein O. A Contribution to the Long Term Treatment of Semen Condition of Androgens Deficiency with Mesterolone. Arzneimittelforschung 16, 466 (1966).

7. Weller O. The Androgenic Activity of Mesterolone in Comparison with 17 a-methyl Testosterone. Arzneimittelforschung 16, 465 (1966).

8. Jost A. La Fonction Endocrine du Testicyle Foetal in: La Fonction Endocrine du Testicule, P 8-18, Masson & Co, Paris 1957.

9. The International Congress of Fertility and Sterility. (Naples 1956).

10. Schirren C. Treatment of Postpuberal Leydig Cell Insufficiency with Mesterolone. Arzneimittelforschung 16:463 (1966).

11. Heller C G, Nelson W O and Roth A A, J Clin Endocr 8, 435-466 (1948).

12. Schirren C. "Neue Ergebnisse der Andrologie (New Results with Androgens)". Springer-Verlag, Berlin-Heidelberg-New York, 1965, p 69.

13. Rausch-Stroomann, J G, Petry R and Hienz H A. The Influence of Mesterolone on Testicular Function. Proc of the Third Meeting of the Int Group Study. Vol 3, 181, 1958, North-Holland Pub Co - Amsterdam.

14. Laschet U and Laschet L. The Effect of Mesterolone on Gonadotrophin and Steroid Excretion. Proceedings of the Third Meeting of the International Study Group for Steroid 3, 169, 1968. North Holland Publishing Co., Amsterdam.

15. Laschet U, Niermann H and Laschet L. The Gonadotrophin Excretion of Fertility - Disturbed Men Before the Beginning and After Cessation of a Mesterolone Treatment. Paper read at the First European Congress of Sterility in Venice, May 1967.

AN ELECTROMYOGRAPHIC STUDY OF THE EXTERNAL URETHRAL SPHINCTER DURING EJACULATION ON PATIENTS WITH EJACULATION TROUBLE SUBSEQUENT TO COLECTOMY AND RECTUM EXTIRPATION DUE TO COLITIS ULCEROSA AND CROHN'S DISEASE

I. Petersén and P. O. Hedlund

Department of Clinical Neurophysiology, the Sahlgren Hospital, University of Gothenburg and St. Erik's Hospital, Stockholm, Sweden

Much interest has been devoted to the part played by the external urethral sphincter in connection with micturition. During the fifties and early sixties, some researchers reported that normal adults were able not only to start but also to interrupt the act of micturition after having been paralysed with succinylcholine or D-tubocurare (Dortenmann & Bauer, 1955; Lapides, Sweet & Lewis, 1957; Scultéty & Abrándy, 1960; Petersén, Kollberg & Dhunér, 1961). Petersén, Stener & Sundin (1970) found that it is quite possible to relax the external urethral sphincter on demand without initiating voiding the bladder even when the subject feels an urge to void after his bladder has been filled. The conclusion to be drawn from these facts is that the striated external urethral sphincter has little or nothing to do with micturition.

The activity of the external urethral sphincter on ejaculation has been studied by Kollberg, Petersén & Stener, 1963, by Petersén & Stener, 1970, and Kadefors & Petersén, 1970. In the course of investigation, the muscle electrical activity was traced by means of coaxial needle electrodes introduced into the striated urethral sphincter on perineal approach. Subsequent to introduction and attachment of the needle to the skin, the patient was left in solitude for the purpose of bringing about ejaculation by masturbation. The muscle electric activity was registered by a Disa EMG analyser connected to a tape recorder. The trial subject experienced no discomfort from the needle electrodes.

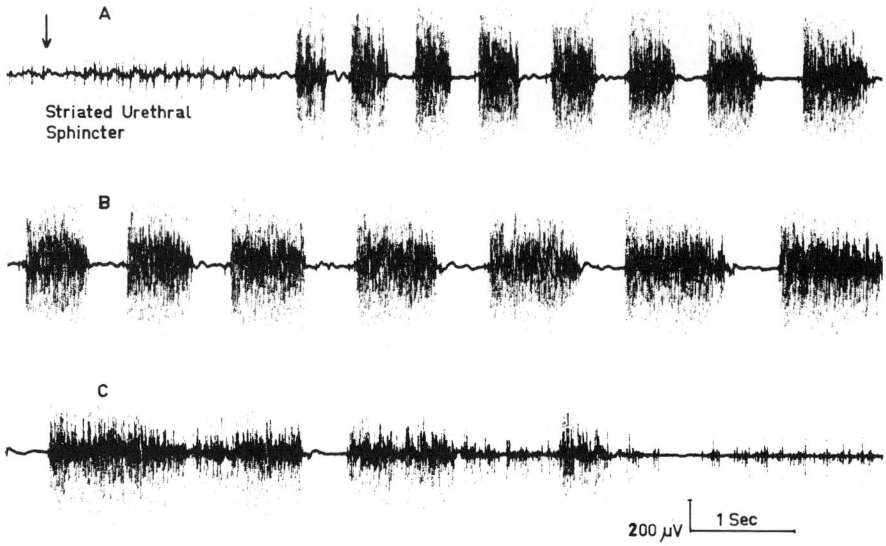

Fig. 1 shows a typical registration of the muscle electric activity from the external striated urethral sphincter on ejaculation. The arrow indicates the moment of the subject signalling that he is becoming aware of the first signs of orgasm. In the 1970 study by Petersén & Stener referred to above, 26 normal volunteers in the ages 25 to 45 years were subjects to investigation. In all the cases, the duration of all periods of muscle electric activity during the contraction of the sphincter, as well as all periods of electrical silence (sphincter relaxtion) were registered. The results of the recordings were treated statistically contraction after contraction, and interval after interval in respect of the entire material. Diagram with regression line and two sigma boundaries for each duration of activity and inactivity (silence), respectively, was produced for the purpose of comparison with the duration values from patients with ejaculation disturbances.

Sexual disorders such as erection insufficiency, orgasm and ejaculation troubles have been described subsequent to operation for canser coli as well as for benign diseases of the colon, such as ulcerative colitis, Crohn's disease and polyposis. Such disorders are ascribed to sympathetic injuries during dissection in the course of operation.

We have studied 8 cases of disordered ejaculation subsequent to colectomy and rectum extirpation. In 6 cases the reason for the operation was ulcerative colitis, and in 2 cases, Crohn's disease. Both capability of erection and sensation of orgasm remained in these patients after operation, but they were unable to ejaculate.

In 7 of the 8 patients of our material, spermatozoa could be traced to a varied extent in the urine sediment after orgasm. The density of spermatozoa in the sediment of urine after orgasm could vary considerably from the same patient on different investigatory occasions. It was constably found, however, that minor quantities only of spermatozoa couldbe traced in urine sediment after orgasm in the case of several of our patients. This would seem to indicate that a reduced volume of semen was expelled in the posterior urethrae from vesicles and vas deferens on orgasm, probably as a result of a sympathetic injury causing a functional disturbance in vas and vesicles.

Electromyographic investigation.

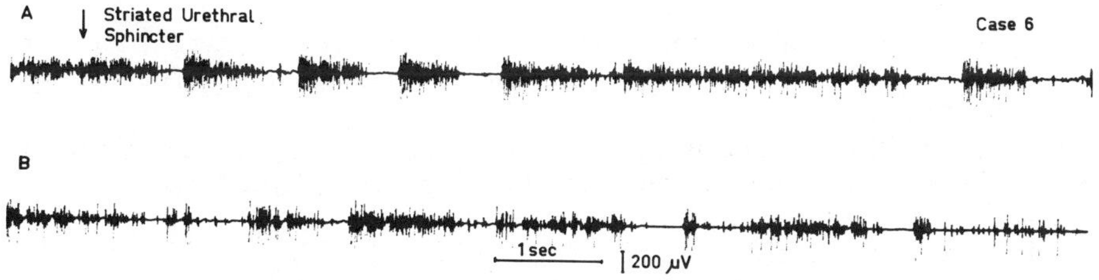

Fig. 2 shows the EMG-record on ejaculation by one of the patients suffering from ejaculation disturbance. The EMG episodes have a very slight amplitude compared to normal extent. The episodes of EMG activity soon become less distinct. As far as the episodes of EMG activity and/or non-activity intervals are concerned, a statistically significant deviation from normal procedure could be observed in 4 out of the 8 cases.

Discussion. The principal reason for the ejaculatory non-occurrence in our patients would appear to be a sympathetic injury having the effect of causing an insufficiency in sphincter urethrae internus, and subsequently ejaculation into the bladder (ejaculatio interna). The function of vas and vesicles may also be weakened in som of the patients on account of the sympathetic injury.

How should the altered EMG record from the external urethral spincter be interpreted in connection with orgasm? One of the possibilities to consider is that the EMG alteration corresponds to a genuinely deranged function of the external urethral sphincter which may be assu-

med as contributary to the disordered ejaculation on account of reduced expulsion capacity. The cause of such deranged function of the external urethral sphincter may be:
1. Injured afferent and efferent nerve-paths to the external urethral sphincter as a result of dissection in the course of operation.
2. Mechanical damage due to shrinkage of the cicatrization from the removed rectum and anus, bringing about positional changes, or possibly restriction of movements of the muscles in the pelvic floor.

Another interpretation is that the changed EMG record from the external urethral sphincter only reflects other functional disorders in the ejaculation mechanism and that the external urethral spincter in reality is intact. Such causes of the EMG change may be:
1. An alteration of the afferent pulse-flow from bladder neck, vesicles and vas deferens on initiation of the orgasm (the emission phase), causing a change in the efferent impulses to the external urethral sphincter in connection with ejaculation.
2. Reduced volume of spermatozoa in the posterior urethrae, resulting in less streching of the muscle fibres in the external urethral sphincter.
3. A change of the afferent impulses on orgasm, bringing about an altered efferent flow to the external urethral sphincter.

A less known fact is the significance which the lack of rectum and anus may have in connection with orgasm, and thus also with the ejaculation mechanism. The fact that a cavernous tissue in the rectum (corpus cavernosum recti) has been described, and that anus is an erotogenic zone, seems to confirm the assumption that there is an interconnection between these structures and orgasm.

An analysis of our method of operation gives rise to the following theory of the origin of the sympathetic injury:

The sympathetic nerve is not injured at its plexus round the lower portion of the abdominal aorta and on bifurcation. In the course of these operations there is no striving for radicality with extirpation of the lymphatic gland, and the dissection therefore does not take place in close proximity to the aorta. Neither will there be any injury of the presacral plexus, since we always pass ventrally of fascia Waldeier on dislodgment of rectum. The sympathetic injury probably occurs on dissection in the pararectal spaces, or on transition from same to vas, vesicles, prostate gland and bladder neck.

It is possible that injury of the motoric nerves to the external urethral sphincter may occur on dissection in the pararectal speces.

References.

Dortenmann, S. & Bauer, K. M.: Untersuchungen über den Einfluss der quergestreiften Muskulatur auf die Blasenfunktion mit Hilfe eines Muskelrelaxans. Medizinische, 1955, 15, 528-532.

Kadefors, R. & Petersén, I.: Spectral analysis of myo-electric signals from muscles of the pelvic floor during voluntary contraction and during reflex contractions connected with ejaculation. Electromyography, 1970, 10, 19-42.

Kollberg, S., Petersén, I. & Stener, I.: Preliminary results of an electromyographic study of ejaculation. Acta chir. Scand., 1962, 123, 478-483.

Lapides, J., Sweet, R. B. & Lewis, L. W.: Role of striated muscle in urination. J. urol., 1957, 77, 247-250.

Petersén, I., Kollberg, S. & Dhunér, K. G.: The effect of the intravenous injection of succinylcholine on micturition; an electromyographic study. Brit. J. urol., 1961, 33, 392-396.

Petersén, I. & Stener, I.: An electromyographic study of the striated urethral sphincter, the striated anal sphincter, and the levator ani muscle during ejaculation. Electromyography, 1970, 10, 43-64.

Petersén, I., Stener, I. & Sundin, T.: An electromyographic and roentgenologic study of micturition, 1970, to be published.

Scultéty, S. & Abrándi, E.: Wirkung der Muskelrelaxantien auf die Blasenfunktion. Z. urol., 1960, 53, 103-109.

INDEX